# WOOD HEAT

Other Books by the Same Author

*The Manual of Practical Homesteading*
*Building Stone Walls*

# WOOD HEAT

By
## John Vivian

**Illustrated By Liz Buell**

**Rodale Press,**
Emmaus, Pennsylvania

PRINTED IN THE UNITED STATES OF AMERICA
*on recycled paper*

**Library of Congress Cataloging in Publication Data**
Vivian, John.
  The new, improved wood heat.
  Published in 1976 under title: Wood heat.
  Bibliography: p.
  Includes index.
  1. Stoves, Wood.  2. Fireplaces.  3. Wood as fuel.  4. Heat-
ing.  I. Title.
TH7437.V58 1978     697'.04     78-17942
ISBN 0-87857-241-4 (Hardcover)
ISBN 0-87857-242-2 (Paperback)

4 6 8 10 9 7 5 3 (Hardcover)
4 6 8 10 9 7 5 3 (Paperback)

For Susa
And all the other woodburners  .  .  .
from Nichewaug and points north.

# Contents

*"I smell yo' bread a-burnin'; turn yo' damper down.*
*If you ain't got a damper, good gal, turn yo' bread around."*

Jimmie Rodgers
"Mule Skinner Blues"
(Blue Yodel Number 8)

# Preface
# to the Second Edition

Though *Wood Heat* has been in print for only three years now, a revised, enlarged and updated edition seems appropriate—indeed essential if the book is to continue as an authoritative sourcebook on use of wood as fuel. When Rodale Press, Liz Buell and I first began the project in the mid-1970's public interest in wood heat was just beginning to flicker. Now, several years' of OPEC-mandated oil price increases, natural gas shortages, coal miners' strikes and increasingly brutal and unpredictable winters have built that flicker to a roaring blaze.

The rapid rekindling of interest in wood as fuel has had both good and bad effects. On the good side, a lot of half-forgotten knowledge (hard-won know-how hauled out back along with the wood stoves once petroleum fuels came along to "save" our great-grandparents) is being unearthed and put to good use once again. In the national imagination, wood has become a hot topic if you'll pardon the poor joke, and entire magazine or newspaper articles, seminars and speeches, academic research papers and scientific tomes are being devoted to the finer points of stacking wood for quickest drying, cooking on a flatiron range top, felling a wrong-leaning cordwood tree and so on. Good stuff, and much of it still lost knowledge (to us) just a few years back. Well, we're cramming the best of the best into the new *Wood Heat*.

On the not-so-good side, a good many of our contemporaries seem to have forgotten that for hundreds of years before chain saws and hydraulic log splitters, every length of wood had to be felled, cut and split, hauled and stacked by hand. As a result, many of the best minds of the day worked on techniques and devices that would squeeze the utmost heat out of that fuel—so that by the 1890s or thereabouts wood heat was a truly high-technology industry.

However, when the '73–'74 oil embargo set off the wood heat revival, few of those eager to take up wood heating and cooking or to capitalize on the new demand for woodburning information and tools bothered to dig up and check out the know-how accumulated by preceding generations. Importers began selling tiny little European "airtight" stoves with a smoke baffle system that buyers quickly found didn't live up to the promises made—heat your whole house, keep a fire overnight; you know the story if you've read any wood stove ads. Worse, every backyard mechanic with a welding outfit began turning out welded steel stoves or fireplace grates made of auto exhaust pipe bent in a series of **C** shapes to draw cool air through the coal bed and circulate more heat into the room. Whatever design caught on was instantly copied; no one knows how many patented air-circulating grates there will be a year hence, and the copies of a single stove idea—the Fisher step-stove with a top bent in a lazy **Z** shape—number in the dozens.

However, just as the small imports were not properly evaluated for use in American homes and our open-space lifestyle, neither were the hot new domestic ideas given a proper testing before sale. Many of the early tubular grates simply burned through. Thin exhaust pipe isn't made to withstand the 2,000° plus temperature of a wood fire. And with use, many of the new stove designs are proving to be bummers. Several step-stove designs are warping; in others, poorly designed or executed welds are splitting. Those are just a few of the hundreds of examples.

So it must be with any new industry. People too anxious to cash in too quick make mistakes on both the buying and selling ends. In the process we all learn—and the past several years have been a real learning period for all of us woodburners. I hope we've included the most important product-related lessons and the best of the newly discovered "old-time" knowledge in the new *Wood Heat*. I can vouch that I've put in some pretty important lessons I've learned personally in the past several years—learned, as usual, the hard way.

In a second edition of a book we have the advantage of several years' constructive criticism. Liz and I would like to thank Bill Hylton of Rodale Press for his editing skill, adroitness at soap-box busting and many ideas for improvements; Bob Ferguson of Kingston, Ontario, and Jim Merriman of Springfield, Massachusetts, for pointing up some discrepancies between text and illustrations in the fireplace section; and my personal thanks go to Woodson Ganaway of Mountain View, Arkansas, for a welding lesson, Robert Cormier of Levant, Maine, for a math lesson and to any number of knowledgeable reviewers and readers for a variety of helpful criticisms, including pointers on how not to treat an ax.

And finally, though we've done our best to put together the most authoritative wood heat guide we can, neither author, illustrator nor publisher can take responsibility for any harm, loss or hassle resulting from installation or operation of your own wood heater even if you follow our suggestions to the letter. What works for us, what satisfies our own building and fire codes may not work for you. Direct, hands-on experience with your own stove or fireplace over several years is the only satisfactory teacher. Till you have it, never forget that just one errant spark can burn your house down. I should know, as you'll find out on the next page.

A couple of years back, still on the farm, Sam and Daddy read the seed catalogs before a fire in the same Franklin that nearly burned the house down. That incident took place long before Sam came along and before Daddy learned what he should have about wood heat. This stove was not set up for the best wood heat; the fire basket is really better for a coal fire. But that's the way she came.

# Introduction

When Louise and I took up serious wood heating some decade and a half ago we nearly burned down our newly acquired antique farmhouse on the first try. Like a lot of people in the turbulent decade of the 1960s, we became fed up with the super-consumer American way of life, among other things, and decided to seek a life as free as possible of the cash economy. During several years of moving from town to suburb to little farm, all within commuting distance of city jobs, we picked up enough skills of rural self-sufficiency that we finally felt ready to cash in and move to a hardscrabble farm in the New England woods.

An early objective was to return the old house, "modernized" with indoor plumbing and kerosene space heaters in the 1920s, to wood heat. We had no more idea than anyone else of the coming energy shortages; the wood was simply growing around us, free for the cutting and hauling. In our years of preparing for the independent life, Louise had become a fair-to-middling open hearth cook and I'd learned to lay and light a grate fire with less than a half-book of matches. I'd read a lot too, and figured I knew about all there was to know about wood heat.

One of our first purchases with our limited savings was a small Franklin-style open-front wood stove. It was a two-door type with a single cook lid on top, patented almost 200 years ago, and we were able to pick it up for just a few dollars back then. We'd found a stovepipe hole already chipped into the antique brick when we opened a flue that must have been plastered up since the Civil War. Fortunately (it turned out), the hole was several inches higher than the stove's smoke outlet. To raise the stove, I built a hearth from what was handy, old firebrick taken from a pile left when another chimney was torn down years ago.

Assuming that the stove would act like fireplaces we'd known in the city, I connected it to the chimney with a short length of stovepipe. Louise put a kettle of cider with cloves and a few cinnamon sticks on top to mull properly. Then we scrounged around the barn for firewood. After packing the stove with dry pine limbs and dusty old maple chunks, we lit a few newspapers under the grate and sat back to enjoy our first fire.

It was lovely at first as the dry pine caught and the bright flames quickly licked higher and higher, filling the dark evening with cheery warmth. The pine limbs did seem to burn unusually fast, but we just tossed on the rest of the old maple chunks and went out to the barn for more. But when we got back, it was obvious that something was wrong. Even out in the entry we sensed an odd odor coming from the house—not the good smoke of a wood fire but a hot, metallic scent more like an overheated foundry. We rushed into the living room to find the fire going with a deep roar. The bed of pitch-filled pine coals was superheating the tinder-dry maple, and the fire was becoming a growing inferno. White hot flame filled the firebox, pouring up and back into the chimney. The length of stovepipe showed a brightening glow, and the cider suddenly erupted into a rolling boil.

With our only fire tool, an undersized pair of tongs, I scrabbled at the stove doors, dodging spattering cider and popping sparks. The stovepipe was almost a cherry red now and the sides of the iron stove itself were beginning to show a dull crimson. I managed to get the doors pulled out and lodged partly closed, but the howling only grew louder and the fire hotter as air was pulled at greater speed through the restricted opening. The cider kettle began to thump on the stove top, and the stove itself, I swear, started to hum and shake as it poured out increasingly intense heat. I want you to know that I was shaking as I realized that our farm, indeed our whole new life, just might become nothing but a pile of charred beams in a cellar hole.

We had picked up a garden hose earlier in the year, and Louise rushed out after it when my attempt to close the stove doors appeared useless. But then, in she rushed, white-faced and more scared looking that I'd ever seen her, and beckoning me out. The night sky appeared alight! Our chimney was belching sparks and little licks of flame and cherry red coals were scattering over the roof. And the hose proved to be too short by half to reach into the house.

With her usual presence of mind Louise began to hose down the chimney and roof while I went back in and, unwittingly, made my only sensible move of the evening. With arms and hands wrapped in wet towels against the heat, I slowly poured the hot cider into the howling flames. Whereas cold water in any quantity could have caused the hot iron to split with near explosive force, the boiling liquid turned to steam on hitting the metal and the draft pulled the fire-quenching vapor into the coals and flames. Slowly the flame subsided and coals dimmed, and finally there was nothing but a steaming bed of black char.

The stove continued to radiate heat, but both the cast iron and sheet metal of the stovepipe quickly lost color, crackling as they contracted in cooling. When I stopped shaking enough to make it outside, I found that the chimney blaze was out also—it turned out to be nothing but flame from last year's chimney swallow nests plastered on partway down the flue rather than the truly serious tar-based chimney fire that has brought down many an old house. We had a proper mess of burned cider and ashes inside, the stove never regained its color along the sides (it looks bleached and ashy after a fire or two no matter how much stove black we put on), and the roof shingles of the house sport a scattering of burnt holes. Thanks mainly to the purely accidental brick hearth, which prevented the stove from burning its way through the pine-board floor and erupting in the cellar, our home was safe.

So through dumb luck and a smiling Providence, our homestead survived my ignorance of wood stoves. You may be sure that after that close call we set out quickly to learn about dampers (I hadn't installed one), draft controls and other fire regulators, as well as cleaning flues, putting in heat shields, judging which wood to use in what quantities in which stage of dryness and at what stage in a fire's life. A fired wood stove is very much a living thing, one that must be tended carefully—or it just may kill you.

Most of this book is based on our own less harrowing experiences with wood heat. Wood was our primary fuel during our near-decade on a farm in the woods. When we traded most aspects of farmlife but the woodlot and garden for a place closer to town over the duration of the kids' school years, we learned how wood can be a town dweller's supplementary fuel—and reduce the gas or oil bill by three-quarters, or more. But wood heat is a broad subject. We don't pretend to have done it all and we welcome expert help. Liz Buell, who did the illustrations, assisted with more than the pictures. She and David and the kids live way up in North Sullivan, Maine, entirely on wood heat. They offered many good tips. We'll introduce other folks who helped out too, as we go along.

To start, let's go into the basic chemistry of wood heat, then the history of man's progression from pre-industrial dependence on wood for fuel, through the heyday of coal, then petroleum and now, for many of us, back to wood. Next we will go into flues and chimneys. (A flue is the long, vertical opening inside the chimney. You might say that a chimney is what surrounds a flue.) Flues aren't terribly exciting, but are the most important part of a wood fire and we'll cover all kinds, permanent types and occasional-use flues that you can build yourself of stovepipe or "cattied" sticks and mud. Permanent flues of brick or cement block should be erected by a mason, but we'll show how you can inspect and repair your own when feasible. I will also give directions on building a cement block chimney you can try if you have the ambition.

Wood stoves for heat are next and we'll discuss the pros and cons of the major types available: Scandinavian and other European imports, U.S.- and Canadian-made, fabricated of iron, steel, sheet metal, firebrick and soapstone. We'll do our best to help you evaluate old stoves, then put them back into working condition. There are photos of restoring an old heater. We'll also picture the job of making a stove from a steel drum, and will install several kinds of stoves, in different typical locations, then get them fired up right. We'll end up with central heat from wood or a wood/oil or wood/gas furnace.

Fireplaces are not as efficient heaters as stoves are, but a lot of homes have them. So we'll try to help you make yours as efficient as possible. I'll admit a bias in favor of a fireplace design you may have heard of—the Rumford, devised by a contemporary of Ben Franklin (who in turn designed the Franklin fireplace stove I'm sure you've heard of). Unlike Ben Franklin, Rumford took the wrong side in the American Revolution; he doesn't get fair treatment in our history books. But his fireplace is a dandy and we'll describe it in enough detail that you can put one in if you're building a new house or extensively remodeling an old one.

Louise is our cooking expert and the following chapter will include all she has to tell about cooking on the open hearth, stoves and wood ranges, plus a lot of information provided by Liz Buell and other wood cookers. (Each wood stove—the oven in a wood range in particular—is an individual. No wood cook can tell you precisely how to operate yours, but their combined experience with their own can get you started.) The paraphernalia associated with wood cooking is fun: trammels and trivets, danglespits and gridirons. And the food is better than anything a modern gas or electric stove ever turned out.

Then we come to wood itself. Many people will be purchasing their supply from a commercial woodcutter, so we'll try to help you find a good one and judge his wood. For those like us who prefer to cut as much of their own wood as they can, we'll touch on finding and evaluating a woodlot, managing it for fuel, and the basics of felling trees, aging and getting in the wood.

Next comes a forthright discussion of perhaps the least pleasant aspect of wood heat, cleaning up. We'll discuss the smoke residue and soot buildup you can expect on every flat surface, and the fact that the old-time custom of spring cleaning comes alive in a wood-heated home. Cutting tools will get their share of attention and we'll rett out the woodshed. In a wood-heated environment you get grubby enough to know the pleasure of getting clean.

Then, to end up there is a bit of speculation about the future of wood heat and a list of selected suppliers of equipment and a bibliography that's as complete as we could make it, so you can check out all the available information on this old but just rediscovered way of life. And that's what wood heat is, really. More than a way to keep warm, it's an expression of a more natural, conserving attitude toward this threatened globe and all who inhabit it. There's even a political side to wood heat and we'll get into that when the opportunity arises.

# CHAPTER ONE

# The Science and History of Wood Heat

Before we get into the details of how to heat and cook with wood, let's check briefly with the chemistry books for an understanding of the way wood is produced and what happens when you burn it. If we take the time to gain a basic understanding of the "why" of the whole process, the "how" will make a lot more sense.

## Wood Heat Chemistry

It is a basic fact of existence on our planet that (nuclear power aside) every smidgin of useful energy at our disposal originated as rays from the sun stored up in winds, rain clouds, tides, or in the case of fossil fuels and wood, by green chlorophyll-bearing plants. The process plants use to capture the sun's energy, called photosynthesis, is as close to magic as we are likely to see. Somehow (scientists can't yet duplicate it artificially in the laboratory) the chlorophyll in plants' green leaves and stems takes the elements contained in two inert compounds, water and carbon dioxide, and uses the energy of sunlight to turn them into a carbohydrate, glucose sugar. The elements are carbon (C), hydrogen (H) and oxygen (O). The chemical formula, if you are interested, is $6(H_2O) + 6(CO_2) + Energy = C_6H_{12}O_6 + 6(O_2)$.

Now this deceptively simple process is the basis of all life as we know it; the plant sugars are the building blocks of all we and other living things eat. In addition, the six molecules of $O_2$ the plants put out along with the sugar provide the oxygen that you and I breathe. But what interests us here is not the food and air that plants offer us, but the energy they pack away while doing it.

Look at the formula again; the energy part you see on the left side of the equal sign doesn't appear on the right side. That

1

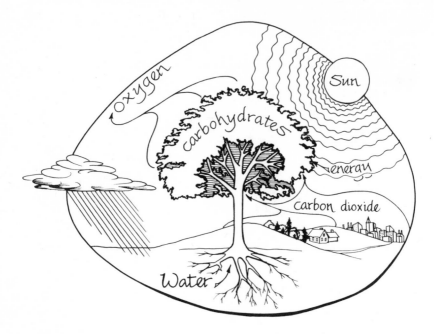

energy is trapped in the glucose sugar. Now, as they grow, many plants—most trees—take some of the sugar they manufacture and turn it into cellulose (glucose sugar is $C_6H_{12}O_6$ while cellulose is $C_6H_{10}O_5$). Not much apparent difference on paper, but by removing a little bit of hydrogen and oxygen, the plant turns sugar into one of the most indestructible living compounds. Cellulose is what makes wood hard. Chopped, cooked and spread thin, cellulose fiber becomes paper—and so tough is it that the fiber can be reused many times. Indeed, the book you are reading now is little more than cellulose fiber—recycled.

Now, there are relatively few things in nature that can harm cellulose. The only creatures that can digest it are a few specialized types of bacteria, including one kind that lives in the innards of termites. The termite feeds the bacteria, and they feed the bug. But even with the help of termites, natural decomposition of a mature hardwood tree takes many years. About the only natural force that can destroy wood in a hurry is fire. Once you start a log burning, it will continue on its own, more or less in a reverse of the photosynthesis process called combustion or rapid oxidation.

During oxidation, free oxygen is added back into the chemical process, turning the wood back into its basic com-

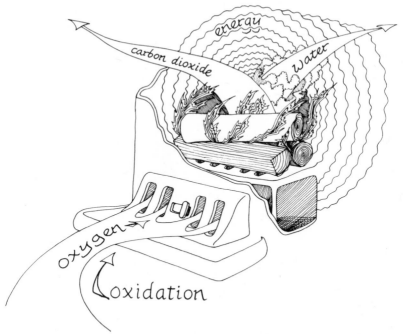

ponents, $CO_2$ and water (plus a few other compounds we'll get into later). Most importantly, the sun energy that went into making the wood is now released as light and heat. Needless to say, the more oxygen you supply a fire, the faster the wood will oxidize—the faster, hotter and brighter it will burn. This is why you blow on a fire or put a bellows to it to get it going.

## Fire Safety

Chemically, a wood fire is nothing but a reversal of nature's growth process, progressing at greater speed. Oxygen supply is the key to the fire's intensity; if we can control the supply of oxygen the fire receives, we are in control of the fire itself. Properly designed and installed wood stoves, fireplaces and flues come equipped with a battery of dampers, draft controls, doors and such that help control the air the fire receives. Learn to use them properly and you are on the way to having a safe fire. The other half of safety involves proper choice of stove, correct placement of stove and pipe, construction and cleaning of flues and fireboxes. But, with any fire, there is always the chance of a mistake, a flawed pipe or fire screen that can let out an errant spark.

Just one spark can burn your house down and you with it. So before you light that first fire, get yourself a good fire extinguisher. If the experience that Louise and I had with our first blaze is any indication, an extinguisher should be a higher priority on your "buy list" than that first stove or set of andirons for the fireplace. And even if it is long enough, don't rely on your garden hose; it may be frozen up or disconnected, or the fire can put your pump wiring out of commission. Don't settle for one of the old-fashioned water or "dry" extinguishers. Get a pressurized, universal extinguisher and hang it so it is handy to your stove or fireplace.

### Fire Extinguishers

The best (universal) extinguishers are rated to extinguish fires of A, B and C class—routine, grease and electrical fires. Plain water will do for most wood fires—Class A. But what if you are frying bacon on the stove when it overheats, then suddenly the bacon grease catches and before you can get ahead of the sputtering, smoking grease fire, it proceeds to fuse the wires in the light socket overhead? A highly unlikely set of coincidences, I'll agree. But it has happened. So, save yourself worry—and maybe save your home, to say nothing of some pretty important lives. Get the biggest and best all-class extinguisher you can find. Better, get two. Put one on the wall between your fire and the nearest exit—and it's a good idea to have your telephone there too. That way, you can clear the house of people, call the fire department and extinguish the fire as you go in. Moreover, the exit is always at your rear, so even if you lose to the fire, you won't lose everything.

Put the other extinguisher somewhere between the stove or fireplace and your sleeping quarters. Many homes have bedrooms on a second story. Put an extinguisher in a handy place on a wall on the upstairs landing so you can grab and go without thinking about it if roused from a deep sleep by smoke.

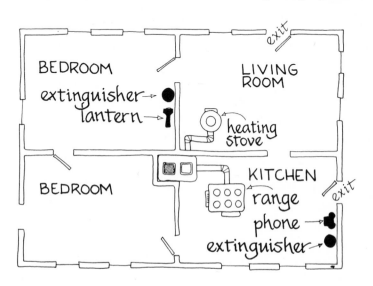

It's best to have a good, well-charged electric lantern in the same spot too. And, needless to say, have the family well rehearsed for a quick escape from the upper story or stories. At our place we can get out on the porch roofs from any number of upstairs windows. Lacking a house so designed, you'd best also put in fire stairs, rope ladder or knotted ropes for fire escapes.

I'm not trying to scare you off wood heat with all these warnings—far from it. But modern home heating and cooking technology has just about removed the danger of appliance-caused home fires. Today's electrical stoves and heating units have circuit breakers, gas ranges have automatic cutoffs that stop gas flow when the fire goes out, and home gas and oil furnaces have a half-dozen fail-safe mechanisms—all designed to protect you without any thought or effort on your own part. Well, fine and good and three cheers for the Home Heat Council.

But with wood heat of any sort, we are getting into an earlier technology, back to the appliances of an era when machines did less for people. Back then, you split your own wood yourself, stoked your own fire and had the know-how to provide for your family's safety. Trouble is, even back in the all-wood-heated 1800s, houses burned down despite generations of built-up caution and knowledge. Today's neophyte wood burner, who like ourselves was probably raised around self-cleaning ovens and all-electric heat, takes on the chore in blissful ignorance of the potential dangers. Or at least Louise and I did, assuming without thinking that a wood stove was just as safe as the complicated oil furnaces we'd had in suburbia. (Of course, we were blissfully ignorant of the complex workings of the oil burner too—still are, for that matter. Something goes wrong, and you just call the serviceman, right?)

Well, there's no serviceman for the wood heater, and no need. In place of all the automatic gadgets that he has been trained to keep going, you'll usually have nothing but a set of manual controls that appear simple, but that can be manipulated in an amazing number of ways, plus your own sense. Or your own senses, I should say. In time you'll be able to tell how things are going by the sounds your equipment makes, by the smells it gives off and the intensity of heat and cooking speed. And you should know the conditions of your fires, or be able to find out quickly, at any hour of day or night.

**Caution**

The secret of safe wood burning, I guess—after knowing precisely what you are doing—is to "heat 'n' sleep scared." You don't have to live terrified by any stretch of the imagination, and certainly shouldn't exist in fear borne of ignorance or uncertainty. But a healthy caution is part of everything about wood heat. In cutting the wood, you keep a sound respect for the angle of the ax stroke, so it won't slip out of the cut and put you on crutches. And before the tree fall gets much underway, you hightail it; not only away from the falling trunk, but also to avoid the butt, which can swivel or split off the stump with great velocity. So it is with the wood fire: retain a knowledgeable respect for it, and it will heat and cook for you faithfully. I can recall many a night Louise or I have been awakened with a start by an odd sound or smell coming from the stoves. To date, it's been nothing but a downdraft from a strong wind gust pushing smoke down the flue or a loud pop as a length of stovepipe cooled down or another harmless incident. That's so far—we continue to stay alert. With wood heat, you have to train your own sense and senses do what technology does for you in a modern home.

## Smoke Detectors

Now, how about those electronic smoke detectors—gadgets that use either an ionization chamber or photoelectric cell to detect the first whiff of smoke or excessive buildup of $CO_2$, then let out an ear-splitting squawk that would wake the soundest sleeper. For years I was against them; they used to cost $50 and more, were less than perfectly reliable, and none can detect a flue fire (the most common wood heat hazard by far) before it's too late. During our years in the woods, I was content to rely on my own senses, sleeping in a room right over the stoves and next to the flue.

Now, we are living in a somewhat more modern, soundproof house, with non-fireproof neighbors not too far distant—a much more typical situation, I'd think—and we've installed a smoke alarm. Sure, the damned thing goes off when Louise burns the toast or I'm generating a lot of smoke trying to warm up a cold flue on an unexpectedly chilly October evening. But in any wood-heated home where you can't carry on a conversation with the equipment all night long, I recommend a smoke alarm.

Ours is located in the upper hall just at the top of the stairwell, so any rising smoke will get it before it gets to us. (Smoke, not flame, is the number one killer in house fires.) People with spread-out homes may want several alarms. The alarm itself will come with installation instructions and most fire departments are more than happy to help you locate an alarm or two. I asked our own firemen for advice on which brand or system to pick and they said to get the battery-powered detector on sale at Sears for less than $20, so we did. It keeps us from burning the toast too often and, I must admit, makes for sound sleeping. Still, we listen to the stoves and sniff the air, asleep or awake. Technology is fine within its limitations. But I still keep alert for a fast crackling in the stovepipe which could mean a draft control carelessly left open, a runaway blaze, pipe or flue fire that I can quench before a real alarm-activating fire gets under way. So, please learn the language of your own wood-burning equipment and consider the mechanical alarm a strictly secondary safety device. The best attitude in my opinion is to install one and then forget you have it, except for annual battery replacement if needed. (In our town you can forget even that.

Get your name on the fire department's list, and they'll call you at battery replacement time. Maybe your firefighters do the same. It's a good municipal service to encourage if they don't.)

You can get alarms that run on house current; they usually contain a little red light that glows constantly, telling you the alarm is in operating condition. Cheaper in the long run I guess than the battery types. Still, I vote for the battery models. A fire could start in your wiring and put the house current, and the alarm, out of commission before it gets a chance to sound.

There's some controversy over which system of detection is best, the ionization chamber that detects by-products of flame before they are visible or the photoelectric cell that picks up visible smoke (though it reacts to very small concentrations). Some people maintain that the ionization chamber is dangerous because it contains a minute amount of radioactive material. Proponents claim that you get more radioactivity standing in sunlight than standing under their machines; all safety labs and regulatory agencies have okayed the devices, so ours doesn't scare me. It's high up on the ceiling in the corner of a hall landing, though, not smack in the main living space.

On the other hand, I've neither heard nor read any thing that convincingly proves that one system is any more sensitive or effective than the other. (In one sense, you'd almost prefer a less sensitive device in a wood-heated home—to negate the burnt-toast effect.) Still and all, I come out with the flat opinion that any reputable brand of home smoke detector is better than none, and I'm fairly sure your fire chief will agree. If he doesn't, take his advice. But, before you put much faith in the thing, test it. Close down the damper on fireplace or stove and see how much smoke buildup is needed to set it off. If our experience is typical (burnt toast to the contrary), you'd have to have quite a flue blockage or actual fire in the home to set off any alarm. So once again, consider the alarm as a backup. Train your own senses to talk and smell the language of wood heat.

## A History of Fuels in America

When the first British settlers reached what is now New England, they were amazed to find the new land almost solidly wooded. The populated areas of their homeland, and of the Dutch lowlands in which many of them had spent several years of exile, had long since been cut over; most central European trees were, and still are, saved for their ornamental value or for building stock. Heat during the eighteenth century in those relatively mild climes was supplied by open fireplaces burning brush bundled into faggots, occasionally from hay or field grasses, and increasingly from peat or soft coal burned in stoves.

## Wood, the First Fuel

But what a blessing all that wood was in the bitter winters of the colonists' new land. They built their homes of huge timbers and thick clapboards or shingles and packed the walls with insulating leaves. They heated with wood fires kept roaring in three, four and more open fireplaces arranged around a great chimney built in the center of the house. Designed on the European model, most heating fireplaces were relatively shallow, the fires built well out on the hearth so as to radiate as much heat as possible into the rooms. Often, though, the kitchen fireplace was large and deep enough to stand or sit in—and this the families often did on the coldest days.

But all flues were open, lacking dampers or any way of stopping the draft, short of stuffing cloth into the opening between fireplace and chimney. And if there was not a good brisk fire in a fireplace, the draft quickly sucked the room clear of warmth. (One reason that rooms in old north country homes are small and have so many doors was to keep warm air in when the fireplace was going, and to keep the room and its drafty fireplace closed off when not in use.)

Back in Britain such chimney designs had been adequate to drive away the chill of the wet and cool winters. The brickwork of the flues would heat up, and together with a small but cheerful fire on the hearth would radiate enough warmth to keep the house comfortable. But not in New England with its subzero temperatures, howling nor'easters and month upon month of snow and ice. Here the fires had to blaze full force, and blaze they did—using tremendous amounts of wood. It is estimated that the original fireplaces were about ten percent efficient; that is, only one-tenth of the heating potential of the wood went to heat the houses and their occupants. The rest went up the chimney. But the forests were full of trees, and the land had to be cleared anyway for crops. So the original forests of New England literally went up in smoke, at the rate of four to five cords per person per season.

Farther south, in the Philadelphia area, many of the settlers were of German origin. In a somewhat less harsh climate they built of brick or stone with chimneys located at the ends of the houses. Indeed, the first farm that Louise and I lived on was in Pennsylvania Dutch country—with two-foot-thick stone walls and a huge walk-in fireplace. One of our neighbors lived in an even more traditional German-style farmhouse. The stable backed directly against the house, so animals and humans could share warmth (and, I presume, fragrances). At one side was a wood storage room with a fireplace built into the wall. Inside the house was a large combination range and heater. The fire was fed from outside in the woodshed. Heat and smoke flowed through serpentine channels in the stove before entering the flue. This one was of brick, though in Europe they were often made of stone, some decorated with porcelain on brick or ceramic tiles. Of the later designs, many had doors attached to the fire opening and draft openings at varying places in the smoke channel, permitting some regulation of air flow and affording a more efficient furnace.

In today's growing literature on wood heat, we're finding more and more articles on these "folded-flue" designs. As of this writing, there's some stir about stove/fireplaces of traditional Lithuanian or Russian design being built in Maine, and I'm sure that many more will come to light over time. People are probably reinventing modern versions of these antiques right now. All I can say is, before firing up an old multi-channel fireplace or stove or before building a new one, be sure you can get a cleaning brush into all the smoke channels.

### Franklin's Fireplace

All the early wood-burning fireplace designs were extravagantly wasteful of wood. Keeping a device such as our Pennsylvania neighbors' fueled was near full-time work for a hired hand. As early as the mid-1700's shortages of fuel wood were cropping up here and there along the East Coast. One result was to stimulate the ever-inventive mind of Philadelphian Benjamin Franklin. He reasoned that the traditional fireplace permitted too much hot air to escape up the chimney. So, he developed one of the first dampers, a sliding door to restrict air flow between fire and flue. One of his concerns was that the closed rooms of his German-descended neighbors were unhealthily stuffy using the outside-fired furnaces. So, one of his most famous inventions, a freestanding fireplace which evolved into today's "Franklin stove," had a cheerful British-style open fire to evacuate room air as the wood burned.

Franklin also observed that wood burned in long tongues, so that much of the flame itself was lost up a conventional chimney. The woodburning process, then as now, has three distinct chronological phases. First, water vapor must be turned to steam and boiled off before the wood can ignite. Second, the volatile oils—turpentine in pine trees, for example—must vaporize and burn. Being much ligher than air, they rise quickly, producing the tongues of flame. Often oils plus water vapor will escape the firebed before they completely burn. This is a waste of useful energy, plus it produces smoke, or as we'll see later the potential fire hazard, creosote. The final stage is combustion of the coals—nearly pure carbon—which burn very hot with little smoke or visible flame.

In his development, which he called the "Pennsylvanian fireplace," Franklin attempted to assure complete combustion by containing the flames and smoke within the firebox as long as possible. There was a false back in the box—a plate or hollow wall halfway back into the stove. The device was exhausted into an otherwise bricked-up fireplace. Flame and smoke had to pass up and around the center baffle, so the wood had more time to burn thoroughly. The cast-iron shell radiated heat into the room, and in some versions air was also drawn up from the cellar to circulate through channels in the stove body itself and emerge, toasty warm, into the room. In these days of renewed interest in wood heat, a number of inventors are applying for patents on similar devices—freestanding fireplaces such as Franklin's and a number of heat-circulating fixtures for new or existing fireplaces. Few are much more than variations on "Poor Richard's" initial innovations, which Franklin never patented himself. He stated that they were for the benefit of all men, so should be free for all to use. Which is fair since some sources claim that Franklin's ideas weren't original at all, but "borrowed" from earlier inventors.

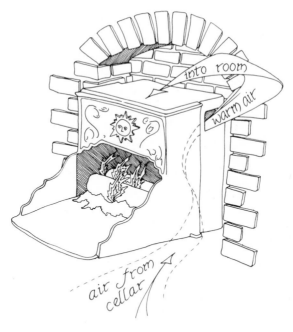

"Pennsylvanian Fireplace"

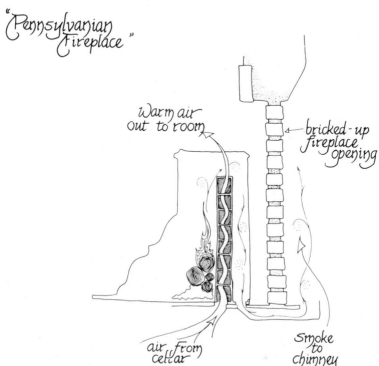

## King Coal

Ben also applied his ingenuity to a result of the diminishing wood supply, a growing dependence on coal. At first it was sea coal, shipped over from Europe at considerable cost. But the high price of imported coal sent the resourceful Americans scrambling for local substitutes. Woodcutters went far afield after charcoal. Acre on acre of trees was cut down, the split logs piled in ricks, then covered and let smoulder. In the process (similar to coking off coal) the water vapor and volatile gases are driven off. The result is nearly pure carbon, thus lighter and more compact. It was less costly to ship the increasing distances between the forests and population centers. But then the Appalachian coal deposits were discovered. Ben Franklin's next heating innovation—a generation after his wood stove—was a coal burner using the "downdraft" principle. He reasoned that if the flow of air through the fire could be reversed—oxygen-laden air coming in from the top—it would draw volatile gases down through the hottest part of the fire, the coals, where they would burn completely. His downdraft stove was heir to sooting up, beaten by the problem it tried to solve, the creosote buildup we'll discuss when we get to the modern "airtight" stoves.

wet leaves →

sod

twigs

Acres of timber...
put into smoldering charcoal mounds.

However, burning updraft or down, coal quickly became king in the rapidly expanding America of the nineteenth century. Coal, of course, is nothing but wood and other vegetative matter that, over the eons, has been compressed to a nearly pure mineral form. Many coal seams bear the imprint of the leafy fronds of the tree ferns that lived during the Carboniferous period, the name literally meaning a time in which sunlight was accumulated, later to be compressed into carbon.

Coal contains more energy, more heat—more stored sunlight—per unit of weight than either wood or charcoal. As such it was far more economical to produce and transport, and with coal, more energy could be stored in a smaller space. Coal-burning locomotives and riverboats could go longer between fuel stops than the woodburners. Centers of population and industry could be located farther from the mines than from the earlier fuel sources, the forests.

When local mining reduced the price of coal, it replaced wood, and the urbanization and industrialization of America began in earnest. At the same time that this power source permitted the concentration of population in towns to serve industry, it permitted the agricultural portion to move onto the relatively treeless Great Plains. It was coal that powered the trains that brought settlers to the West in meaningful numbers. By the turn of this century it was cooking the meals and heating the houses in Kansas City and points west. Wood was too valuable to burn: it went to build the homes and churches and commercial buildings of a growing America.

### The Reign of Oil

Coal remained the dominant energy source for a relatively brief time. Though available in our country in great quantity, it was soon replaced by an even more concentrated form of sun energy, petroleum. Being liquid, oil is even easier to obtain and transport than coal. And it is comparatively easy to break down into even more highly concentrated energy forms such as fuel oil for your home furnace and gasoline for your car. And just as coal permitted easier transport and storage of energy than wood, petroleum products permitted still more flexibility in industry and life styles. Coal or wood could never power a jet bomber, a submarine or modern automobile. The sheer bulk and weight of these solid fuels would prohibit cities in temperate climates from growing to the size of a New York or Los Angeles. Without oil, we never would have reached today's and tomorrow's levels of human overpopulation; both the chemical fertilizers and pesticides on which modern superproductive agriculture is based come from petroleum or its companion resource, natural gas. Which is to say that the discovery of oil has been a mixed blessing.

Further, during the heyday of cheap oil, we've let our other energy resources languish. The coal mines have been let go to ruin and until just the last few years the most visible use for wood was in veneer to panel dens and recreation rooms in which the fireplace was an ornament, the blaze mainly for entertainment.

## Wood Heat and the Environment

Well, those days are gone, and unless the still questionable claims of the proponents of nuclear, solar and other high-technology energy sources come about, they are gone for good. Petroleum is in limited supply and is bound to run out eventually—hopefully not before the powers that be realize that it is too valuable to burn and is better saved to make plastics, medicines and such. Coal is still with us in great amounts, but available only through deep or strip mining, the former costly in money and miners' well-being, the latter in money and environmental damage. And from whatever source, coal releases sulfur and other pollutants into the air. Both these fossil energy sources—these wonderfully concentrated forms of stored sun energy—will remain available for many years, though at ever increasing cost. Which brings us to the main point, the economics of wood as a fuel. As of this writing, in the depth of winter 1977/78, wood is easily the most economical energy source for perhaps one-third to a half of America's families, those living in rural, small town and suburban areas near forested areas. Let's examine the facts and figures.

First, how about air and other environmental pollution; wood fires smoke, and smoke pollutes, right? Well, yes, but it's as close to "good," natural pollution as there is. A wood fire is nothing but a speeded-up version of natural decay, as previously noted. In the forest, every tree will eventually fall, and as it

*Natural decay*

*water*

*carbon dioxide*

*minerals*

moulders away it will release water and $CO_2$ into the air, leaving its minerals in the soil; it's the same process as having a grate fire and putting the ashes in your garden; it just takes longer in nature.

You might suspect that intensive harvest of a woodlot would increase the amounts of by-products of wood decomposition that are released, on the assumption that burning destroys faster than the natural process would. This too is a misconception.

If we operate our own woodlots to have a continual sustained yield, a given amount of acreage will produce just so much vegetation per year. No more than the naturally received sun permits. You can clearcut, removing all the wood in one swoop, then wait 25 years or more for the next cutting. Or you can harvest about a cord of wood per acre every year forever. Either way you are removing only the amount of wood—the stored energy—the land would turn over naturally in the given time period. If you or I don't harvest the wood using selective forestry techniques to produce the best lumber and firewood of the desired tree species, nature will harvest the same amount using her own methods of selection. Clearcutting destroys the beauty of a forest and if improperly done can produce severe land erosion and other problems. But not even that unlovely technique can make the woodlands produce more wood than they would on their own hook. (Restoration of the minerals to the woodlands may be a problem over time, though trees "mine" minerals from deep underground; more research is needed.) However, of our fuels, wood is the only one that is self-renewing, the only one that is nonpolluting. Put another way, burning wood that's harvested sensibly produces no more environmental effect than would the natural decay process.

That's assuming you and I and all the other woodburners apply the same sustained harvest principles to our own woodlots that the big lumber companies do. I have more than one quarrel with the plywood and paper makers, but one thing they do is keep that wood coming on. If we permit too much more of America to be deforested and turned into urban sprawl, our woodlands could quickly become less than self-sustaining. It would take more than Barry Commoner and the whole environmental movement to save the forests if very many people were going cold in their homes. And if you want examples of what happens when the forest is overcut, the world is full of them. For starters China, India, Haiti and much of Africa are treeless by and large, due to centuries of too many people living among too few trees. It's still debatable just how much of the world's deserts are products of climatic change, and how much is due to overcutting by man and overgrazing by his herd animals. But until the nation of Israel began the job of replanting them on their former hillsides—gone to desert—the cedars of Lebanon were only a part of biblical history. I wonder if there would be anyone left to replant the redwoods of California or maples of New England if they were cut out? There are plenty of woodlands now to warm us. Let's all stay politically alert, and keep American green—as well as warm.

## The Economics of Wood Heat

If we do lose our woodlands, it will result from someone's selfish and shortsighted effort to make a quick profit. So, we'll examine the dollars and cents (sense?) of wood heat. First, the larger picture of capital investment and jobs for people from our several energy sources: each gallon of fuel oil or ton of coal you burn is a result of an investment of millions of dollars in refineries and mines, crude oil tankers and coal trains. Practically no labor is required in oil production after the well comes in, and the distribution from wellhead to refinery to your home tank produces relatively few jobs. The main economic result of oil-domination of our energy system has been to make very few men and institutions very rich. Today most of them are in the Middle East, though a few Texans are still doing okay, I hear.

Coal isn't much better. It produces more jobs, but the hard work and dangers of a miner's life plus the dust, grit and grime of the entire delivery process don't make the sort of jobs that excite much envy. Unlike oil, the wealth of much deep-mined coal stays closer to home. But the mine operator's house still occupies the top of the hill; the miners live down in the hollows. The economics of strip mining are more similar to oil production. Huge machines roar in, remove the soil, then gobble the coal seams and move on. Perhaps a few local heavy equipment operators have good jobs for a while. But most of the money goes to the huge company that owns the equipment and mineral leases—more rich folks getting richer.

Now, I'm not necessarily going on like this to criticize the oil or coal companies. Though, like yourself I'd bet, I do enjoy reading about the criticism they are getting from other quarters. And I'm not grumbling about America's highly successful economic system or the folks who accumulate the wealth it produces. Louise and I had our shot at that life; we didn't think the rewards were worth the effort and chucked it in large part. But that's personal preference, not a condemnation of anybody or anything. I'd suggest that switching from oil or whatever to wood heat actually supports a healthy capitalism and wealth accumulation. Not by some distant millionaire, but by the hard-working local guy who needs the money badly enough to go out into the woods and work up firewood. Of course, ideally that person is yourself; property taxes aside, the "price" of wood harvested from our own woodlot costs Louise and me only the price of a little gasoline for the machinery plus a lot of sweat. As Henry Thoreau said, "Wood heats you twice, once when you cut it and again when you burn it."

But even if you purchase wood, most of the cash goes directly to a hard-working crew as pay for time spent. Getting in

One Cord Wood

200 Gallons Oil

firewood is almost entirely labor. The basic capital required is for a truck, a saw and a splitting maul. And woodcutting is a one- or two-man enterprise. Exxon isn't interested—yet. It just seems to me that in these perilous times, your own hard-earned dollars are better spent paying for the honest work of a neighbor than heading off to Chicago or Dallas or Addis Ababa or wherever to fatten the bank accounts of people who are already driving Rolls Royces. Wood heat as global politics? Well, it's worth thinking about.

**Personal Economics**

So much for the big picture. How about the smaller one featuring your pocketbook and mine? Quite simply, in terms of

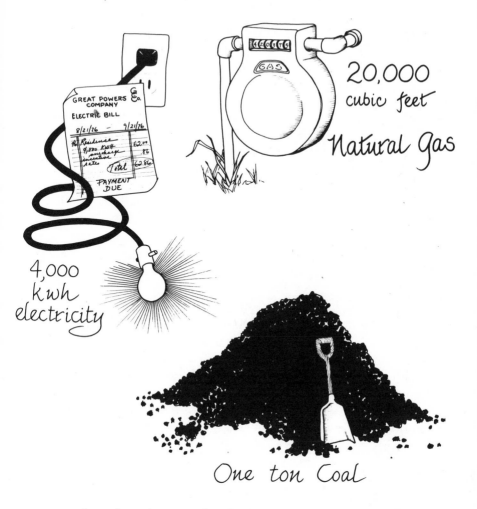

4,000 kwh electricity

20,000 cubic feet Natural Gas

GREAT POWERS
COMPANY
ELECTRIC BILL

One ton Coal

strictly cash outlay, wood is the best energy bargain available if you live reasonably near the woods so that transportation costs are minimal. There is a great deal of fetching and hauling to do, however, and if you put a great premium on your free time, or if your job is such that you want home time to be mainly relaxation, wood heat may not be for you. Let me try to put some rough dollar figures on it, though, based entirely on our own preferences, experience and prejudices.

First, let's try to make a comparison of costs versus heating value for the major energy sources. I've read several books and articles that attempt to equate costs of wood versus oil versus electric heat down to the penny. This is silly because of the tremendous variation in costs of each fuel in different sections of the nation, variations in heat value between different grades of each fuel and the true efficiency of heating plants. For a rough set of numbers applicable to our part of New England during the ghastly winters of the late 1970's see the chart "Comparison of Heating Fuel Costs." I've made an attempt to compute in general terms the amount and costs of commonly available fuels that would put out the same useful heat (stress there on *useful*) as a cord of top-grade hardwood. A cord made up largely of beech or oak weighs some two tons and costs $50—that's as of this writing; demand is bound to push the cost higher in winters to come. But the cost for equivalent heating fuels is around $80 for coal and gas (if you live in a town with a natural gas works— propane or bottled gas would cost four times as much), almost a hundred dollars for oil and over a hundred and a quarter for electric heat.

Now, that's here in our country town. In any city the sheer logistics of hauling wood in, storing it and removing ashes would put the cost at or above costs for oil or gas, that is if you buy your wood. On the other hand, if you spend weekends scrounging around the local state park for deadwood and windfalls and use a little wood stove to keep the reading room warm at night when the kids are asleep and the thermostat turned down, your time and price of a small chain saw will be amply rewarded in saved fossil fuel costs no matter where you live.

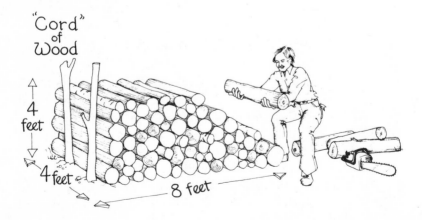

"Cord" of Wood

4 feet

4 feet

8 feet

## COMPARISON OF HEATING FUEL COSTS

| A | B | C | D Gross Btu* Output in millions | E Efficiency of average heater | F Net Btu's in millions (E x D) | G | H Equivalent useful heat cost (B x G) |
|---|---|---|---|---|---|---|---|
| Fuel | Amount | Weight | | | | Cost | |
| Dry Beech Wood | 1 cord | 2 tons | 28 | 50% | 14 | $50/cord | $50 |
| Anthracite coal | 1 ton | 1 ton | 23 | 60% | 14 | $80/ton | $80 |
| #2 Fuel Oil | 200 gallons | 1,500 pounds | 22 | 65% | 14 | 48¢/gal | $96 |
| Electric Service | 4,000 kwh | 0 | 14 | 100% | 14 | 3¢/kwh | $120 |
| Natural Gas | 20,000 Cu. ft. | 0 | 20 | 70% | 14 | 4/10¢/Cu. ft. | $80 |

Note: The above figures are based on average prices for central New England during the middle of the 1977–78 heating season. So-called "fuel adjustment charges" in electric and natural gas prices are included. For contrast, equivalent costs of the other two readily available fuel sources, propane or "bottled gas" and Cannel or fireplace coal would exceed $300. And, though exact figures are impossible to figure, I calculate that the *monthly* cost of a solar heating unit—payment of mortgage, taxes and maintenance—would exceed that $300, and the bill would come in 12 months a year, for our house and locale, at least. Though solar's day has gotta come, it's not here yet in dollars and cents terms.
*Btu—British Thermal Unit

Now, what about the economics of bringing a really substantial amount of wood—say, the five to six cords needed to heat our house and cook our meals over a winter? Your standard cord is a stack of four-foot-long sticks piled up in a rectangular box shape that is eight feet long and four feet in height. A cord of good burning hardwood, even when well air-dried will weigh about two tons. A cord of soft pine will weigh in at perhaps a ton—and, as you'd expect, will last half as long and produce half as much energy as hardwoods. Now to get a cord of any wood you must cut down a tree, saw it up, then split the larger pieces. Back when they had nothing but ax, saw and splitting maul, they figured that a good man could put up two cords a day—from tree to drying stack. I guess those were hardier days. The best I've seen in this modern age was two men, working at a steady pace with chain saws and a four-wheel-drive truck and power winch, getting five cords cut during a good day. The larger sticks still had to be sawed to stove length, 12 inches, which adds in time to put three more cuts along the entire face of the cord. Plus, splitting and hauling were left till later, a good way to get a bit of exercise each evening during a sedentary winter.

Let's see now, those five cords would cost $250 if purchased today. Figure, what with breaks and a long lunch hour spent hunting pheasant in the cornfield, our woodcutters got in an honest five hours of work per man. That would make the "wage" come out to $25 an hour. Not bad. Even if we admit that another day's work would go into cleaning up and chipping the slash as well as stacking and splitting over the winter, the pay still comes out to $12 an hour. Of course, we could value their time in terms of cash not spent for the alternative fuel, heating oil. That would bring us back up to the $25 figure and more. Then we can add on the money not spent in an alternative form of physical exercise or recreation. What does it cost per playing hour to belong to a golf club, $5? That's a wage of 25 dollars each 60 minutes. If we cut wood instead of renting a light plane, the imputed "wage" could jump to $50 an hour and more. What recreational exercise do you pay to enjoy?

But let's never forget that the woodlands will remain self-sustaining only so long as we don't exact too much of a "wage," imputed or otherwise from them. I'll say it once more—let us woodburners insist the loudest that we all keep America green.

# CHAPTER TWO

# Chimneys and Flues, Old and New

The least interesting part of your wood fire is the chimney. However, it is probably the single most neglected or ignored. So, let's start with that all-important tube that keeps the fire burning and your room clear of smoke. When a wood fire burns it generates heat and water vapor and several gases, mainly carbon dioxide. If the fire is young or the wood green, combustion will be incomplete and unburned carbon and assorted oils and resins will come off as smoke. Of course the fire consumes oxygen as it burns, so a fire just let burn in a closed space can make it uncomfortable, dangerous and even fatal for us oxygen-breathing creatures. Shortly after he discovered fire, early man must have found ways to get the smoke out and fresh air into his cave or animal-skin tent or whatever. Hot air rose then as now, so ways were devised to guide it up and out of the dwelling.

Probably the most familiar example of a so-called "primitive" smoke pipe is the tipi of the American plains Indians. These nomadic tribes followed the game in their seasonal migrations, erecting the cone-shaped shelters of buffalo or deerhide on frameworks of fresh saplings at each campsite. The structure had a closable entrance flap at ground level and one or more adjustable flaps at the top to regulate heat and smoke flow. In cool weather all flaps would be opened and a small fire built in a stone-lined pit, normally in the center of the tipi. Like most fires it would be smoky at first, but as it burned down and generated a bed of coals to assure quick and complete combustion of the wood, smoking diminished and the flaps were closed till only enough air came in to keep the fire's gases moving upward so that smoke and fumes accumulated in the high peak of the cone.

Entrance flaps could be closed fairly tight and the top flaps were arranged as a sort of chimney extension—aiming down-

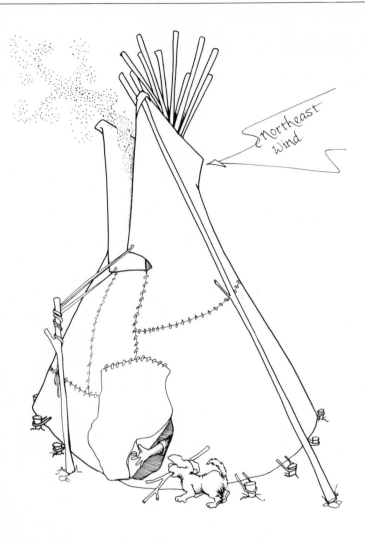

wind and blocking the smoke hole itself from the direct path of
the wind. In this way a smoke eddy and a slight vacuum were
created at the flap—which acted very much as a bird or airplane's wing does in generating the vacuum that lifts its weight
into the air. This effect kept the smoke and hot gases flowing
more or less continually out the smoke hole. At the same time,
fresh air was being drawn in at the hole's edge to replace what
was escaping at the center and along the flap. This cooler air im-
mediately fell, but by the time it reached the occupants, it had
warmed from contact with the tipi sides and the column of hot

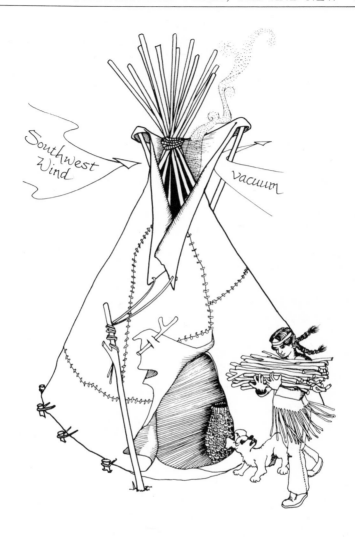

air rising through the center. The Indians were further warmed from heat absorbed and distributed by the earth floor of the shelter.

The tipi, with its relative efficiency in evacuating smoke from the dwelling place is quite a sophisticated design—developed I'd bet by a Stone Age genius with no book learning, but an intuitive knowledge of aerodynamics and thermodynamics. The more typical shelter of preliterate peoples in chilly climates is a dome-shaped structure of animal-skin over a pole framework or a similarly shaped hut of whatever building material at hand—mud and brush or stones; the Eskimos used ice. But by all reports, the smoke from the central fire was evacuated through nothing more than a hole in the roof, and the interior of the shelter got pretty smoky.

Some early peoples came up with wood-heated devices that wouldn't smoke you out of the place, but all the ones I know of were for cooking. The Mexican oven, still in use in one form or another throughout much of the less developed world, is little more than a dome-shaped structure of adobe or mud and rock with a rock or brick door. Some types have a smoke hole in the back. They are heated by being filled with hot coals or having a fire built in the mouth. When the oven is ready, the coals are raked out and the bread is put in on the floor where it is baked by heat absorbed by the oven walls and floor. A somewhat more flexible cooker is the Chinese oven, presumably an early development from China. The idea became mildly popular in the United States back in the 1940s when backyard barbecuing

Mexican Oven

...before bread goes in...

coals are raked out...

Chinese Oven

was first becoming widespread, and you occasionally see one in the backyard of a suburban house. (The oven was often used as an incinerator before air pollution became a severe problem.)

The Chinese oven is nothing but a low chimney with a door at the bottom and a steel lid that has a hook welded on its lower side at the top. Size varies, but waist high is most common. You build a wood fire in the bottom of the device, regulating draft by opening or closing the bottom door (or by regulating a draft control if one is built into the steel lid). Once the walls of the chimney are hot, the fire is let go out or die to a bed of hot coals, depending on what is to be cooked. A haunch of meat, several fowl or a pot of stew is hung from the hook on the lid bottom, the lid goes on, and the meal is cooked by a combination of absorbed heat as in the Mexican oven, plus upward radiating heat and smoke from the coals. It is a full-time job to keep heat constant when cooking with this device, but the result has the flavor and

First, a hole ...          Then, a hood...

juices sealed in by the surrounding heat and the coals give the food a wonderful smoky tang. (We've never built a Chinese oven from scratch but have gotten much the same effect by hanging a piece of meat in the preheated flue of an outdoor fireplace. It is a great way to cook a leg of lamb or a ham while finishing off the maple syrup—boiled all day over the outdoor fireplace but cooked the last crucial hour or so on the range inside. Potatoes can be baked just by putting them in the bed of ashes—covering with foil if you're finicky about a bit of wood ash on the plate.)

## The Development of the Flue

As the human society changed from nomadic or seminomadic hunting or from a foraging orientation to an agricultural one, life became more settled. Villages became towns, then cities. Better and better ways were developed to guide the smoke from heating and cooking fires out of living space. First, homes became more substantial, built of wood or stone, thus less drafty and more capable of retaining heat. It was probably in the relatively balmy climes of southern Europe—Italy most likely—that the fire was first moved from room center to a far stone wall. A smoke hole was put in the wall just above the flame. The wood not only heated the room air by radiation; warmth was absorbed by the walls, which continued to emit heat long after the flames were out.

Next step was to construct a hood on the wall to gather smoke; then a pair of sides was added, extending from the

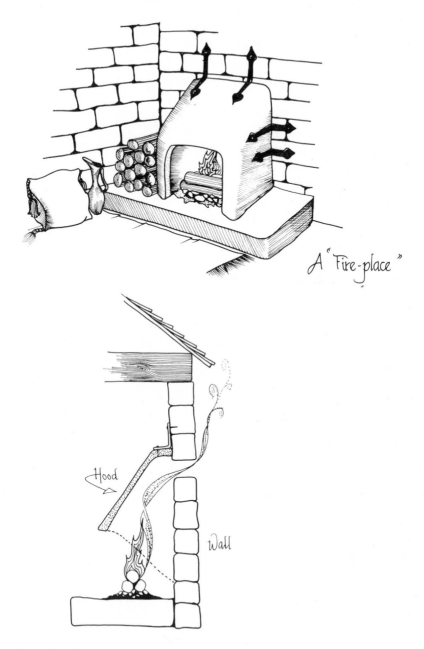

A "Fire-place"

Hood

Wall

hood's edges down to near floor level—all to contain the smoke and guide it up to the smoke hole. The next idea was to further contain smoke by recessing the fire into the wall itself. Thus was born the fireplace as we know it. But somewhere along the line came the major breakthrough—the flue or chimney, the latter term originally referring to the entire apparatus from fire base on the floor to the top of the smoke outlet. The full chimney was a feature of the homes of the affluent British by the twelfth century.

With a flue or not, hot air rises, and smoke from the heat-generating fire is carried along with it. The vapor must stay hot to keep going and the hotter it remains, the faster it will rise. In the open center-of-room fire, smoke spread out along the ceiling and cooled greatly before it got to the smoke hole. Hoods and such did little more than bottle up most of the smoke till it found its way out. But the modern flue actually serves as a smoke guide, an active participant in the fire, which clears the room while pulling in air to supply oxygen to the blaze.

## Flue Construction Principles

In your own wood heating rig, whether you are building a new home around stoves or fireplace, putting an old chimney into really efficient heating operation or adding a new flue to your existing home, there are a few principles you should keep in mind. First, of course, the installation must be safe; we'll get into the details of safety in construction, cleaning and maintenance a bit later when we discuss the various construction materials and techniques. But safety should always be your cardinal principle in *anything* to do with wood heat.

Next in importance in flue construction is flue efficiency. The flue itself should be located so it can go as straight up as possible—this is the *flue*, now, not necessarily the length of stovepipe you use to hook up a wood range or heater to the flue. But once you get the smoke into the chimney, you want it to flow upward unimpeded. This is not only to guarantee optimum draft, but to reduce creosote buildup; get that smoke out of the flue as fast as possible. For the same reason you also want the interior of the flue to be as smooth as you can get it (there are more details on this aspect following).

If you are working with an existing flue, you don't have much choice as to its location. But if installing a new one, locate it inside your house if at all possible. Not only will a flue inside a warm house fire up more quickly; once warmed it will radiate heat into living space, increasing the heating efficiency of your unit. I know that most homes built in the last fifty years or so have fireplaces and flues at the back or side walls of the building. But that's for the convenience of the builder—besides, most modern fireplaces are purely recreational and not meant to heat anything. If you're serious about heating with wood, keep as much of that heat energy inside as you can.

Traditionally, chimneys are made of brick, occasionally of stone, and these days of concrete or cinder blocks. Most building codes specify (and even if they don't, you should incorporate anyway) a ceramic liner inside the outer shell. In general, a brick chimney will last the longest, look best to most conventional observers and cost the most to put up. Cement block chimneys go up fast and are in my opinion the only type of permanent, ceramic flue that should be attempted by a non-mason—details on that later also. We've personally had bad experience with fieldstone chimneys. Brick and concrete block are hard, but still porous so that mortar adheres well to them. Rocks, or the hard, igneous rocks that the glaciers deposited all over New England at least, are smooth and impervious to mortar, so that a rock flue is more an aggregate than a monolithic, well-bonded structure. If put up by a stone mason, now, who knows how to lay a dry (mortarless) stone structure so that it will last, a rock chimney should hold up a while. The mortar is more to fill in the cracks than hold the chimney up. The one I had to deal with some years ago was put up by an amateur and whenever a chunk of mortar was pried out by winter freezes, a rock or two would come with it.

Unless you are building a new house or doing an extensive remodeling job, you'll pretty much have to locate a masonry flue on the outside of the house, with the stove or fireplace against the outside wall. The layout of your house will likely dictate the location of the woodburner, thus the flue. But if you possibly can, locate it downwind—on a wall away from the direction of the prevailing wind. Keep clear of tall trees and neighboring buildings where possible; they can cause eddies and irregularities in airflow across the flue and your roof that can ruin your draft, particularly on a stormy or windy day.

You've a lot more flexibility all around if you elect to put in one of the metal, super-insulated flues that have come on the market in the past few years. They don't weigh the many tons of a masonry flue, so they don't need any special support. They are as safe (or so say all the testing labs, building code agencies and such) as a good brick flue, and at about $1 an inch, cost perhaps a bit more than a concrete block flue but less than a brick one. Plus, you can install one anywhere in your home so long as you can hide, disguise or ignore the run of pipe up through any rooms above the heating unit. We'll get into more detail on installing these flues, too, later on. For now let's begin with the situation that Louise and I have encountered most frequently and found most difficult—reactivating old flues that may have been boarded or plastered up for years. If they've been protected from water damage many old flues are basically sound; ours have been and I hope any you may be putting back into service are, too. Here's how to tell, and how to fix any problems.

## Evaluating Existing Flues

Start up on the roof; that is where most chimney damage takes place and where nearly all of it originates. If your house is an old center-chimney colonial you may find as many as five flues in the single chimney. The great brick structure warmed up and radiated heat through the entire house during the winter. Also the early Colonial governments levied a chimney tax, so the crafty Yankees put all their flues into a single chimney. The number of openings in the chimney top will tell you how many fireplaces there are or were. In one of our houses, the fireplaces had been torn out and a simple buttress put under the flue when the more "modern" and efficient cast-iron stove became popular. Many houses we know have fireplaces bricked up, with only a stovepipe running into the flue; still other fireplaces have been entirely sealed away behind the walls.

South of New England the flues will often be hidden in the walls of brick or stone houses, sometimes with the chimney top removed and roofed over. Here you'll have to pound around with a mallet and listen for the hollow sound that betrays a flue. If your older house has had its flues sealed to be put out of commission permanently, you may be better off building new ones. I'd call a good mason if repairs or rebuilding flues in an old house entailed much more than a bit of patching. Neither Louise or I is an expert on house construction, and unless you are, I'd advise calling in someone who knows the subject cold. Don't put in something that could be dangerous or damage the appearance and value of your house.

As early as Ben Franklin's time, houses were being built specifically for stove heat, and central steam heat was available for homes well before 1900—indeed, had been in use in public buildings as early as the mid-1700s. Here, the flue or flues can

a Colonial Center Chimney

a flue . . .    for every fireplace

a single, central flue...

...for many stoves

be located anywhere in the house—back, sides or through in-side partitions. The place we live in now was built in the late 1800s and is fairly typical of its era. There is a single central flue and into it exhausted a big coal-fired hot water boiler plus three wood stoves: the kitchen range and bedroom stoves on the second and third floors. The coal furnace was replaced with an oil burner some time back. We hope to replace that with a wood/oil burner one of these days (the oil as a self-tending backup for when we are away in the winter).

**Flue Capacities**

If your flue is used with any sort of central heating device, most experts will advise you to build a separate flue for wood heat. It's true that a fireplace and modern furnace should not operate on the same flue. The great amount of draft a fireplace requires can interfere with the furnace's operation. But an effi-

old flue
ready
to re-use

...or it
may
be hidden under
the wallpaper...

cient stove or stoves will not—particularly if they put out enough heat that the central unit does not operate at the same time they are on. More on that later.

If you find flues with no apparent fireplaces and no connection to a central furnace, you likely have a chimney that was built for wood or coal stoves. You may find sheet metal covers over the holes in the walls—many had decorative scenes printed on them. Plates may be buried under wallpaper, or the holes may have been plugged, plastered or cemented shut. Get out the mallet and pound walls. The openings will usually be somewhere near chairback height or higher.

### The Code

Now, before you plug a new stove into an existing smoke hole or reactivate an old fireplace, bone up on your local building codes. Building code details and the stringency with which they are enforced vary from municipality to municipality. It's hard to generalize, but in the main code restrictions cover new building or major renovations that involve movement, removal or addition to a major building component such as an interior or exterior wall. Building a new flue or installing a large stove is usually covered. Many areas honor the "grandfather clause" that permits use of non-code devices built before the code was adopted. Re-installing a stove where there has been one before may fall under this rule in more lenient jurisdictions.

Now, I've got more than one bone to pick with building codes. They are developed by and for builders, trade unions and bureaucrats and in many areas effectively prohibit a person from building his or her own home or doing any home repair or maintenance beyond replacing a faucet washer or two. However, if you don't have a required building permit or put in a non-code installation, you can get into trouble with local authorities. Worse, your home fire insurance may be invalid. Indeed, at this writing and in our town, several liability insurance companies are running around in a panic, threatening to cancel insurance policies for anyone having a wood stove that isn't state approved.

We may chuckle at the pettifogging of insurance actuaries or building inspectors, but the safety of your home and family is no joke. So double-check the condition of any old flue before putting it into operation. Inside the house, remove cover plates over pipe holes, clean out any old fireplaces and poke around in cellar or attic to see what sort of problems you may have. We've found that the old-timers moved interior walls around frequently, and you find flues boxed in in any number of ways with the connectors between flue and an inside room wall made of flimsy stovepipe, half-rotten plaster and I don't know what all. So, be sure you know your flue, what holes may have been knocked through it and what has been done to them. For example, I saw one old upstairs bedroom stovepipe hole in a flue that had been abandoned; the hole was stuffed with an old pillow, the opening on the inner wall sealed up with newspapers and whitewash, then the whole covered with wallpaper. A sure firetrap if anyone had put a stove into a smoke hole on the lower floors.

### Evaluating the Masonry

Now that we know what you have in an old flue, let's get back on the the roof and finish the inspection. There should be a good cap around the flue, a solid mortar coating that prevents water from getting into the top joints and pushing out the mortar. Use a screwdriver to probe around in the mortar joints up and down the chimney. If you can dig out small chunks or even a brick or two, it's alright. They can be replaced. If all the joints are rotten, and particularly if the bricks are crumbly, you'd best plan to rebuild at least the outer layer of the chimney top. Look inside; if the inside shows rotten brickwork, you'd

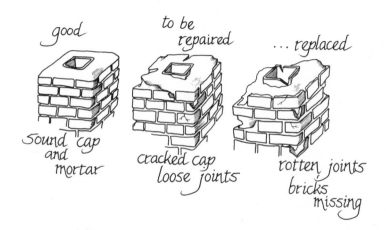

good

to be repaired

... replaced

Sound cap and mortar

cracked cap loose joints

rotten joints bricks missing

best plan to have the chimney rebuilt from top to at least a foot below roof level. If a flashlight shows loose and crumbly joints and broken or protruding bricks well down into the flue, I'd call a mason and be prepared to totally rebuild. The last central chimney (three fireplaces included) rebuilding job I witnessed cost about $3,000 and that was a good many inflation-ridden years ago.

Assuming you can go ahead and use the chimney, check where it mates with the roof. This is a most critical part of the structure, where rainwater is kept out of the flue and the attic. There should be flashing: sheets of lead in older houses, copper in younger ones. One edge of the strip should be embedded in mortar joints or mortared directly to the masonry. Then it should run down to the roof sheathing, and out for several inches under the shingles or other roof covering. The whole joint may be covered with a thick coating of tar. If the flashing is missing or in bad repair, get ready to fix it.

Now, repeat the inspection of joints and brick, stone or whatever on down to the foundation. Get into the attic, cellar and any other locations of access. Look for loose mortar as on the roof. If you see dark streaks that look as though sooty water has flowed out through the bricks, you've got trouble. The problem may be lack of a sound cap; the bricks literally soak up water if you let them. Or it may be rotten joints. Your screwdriver will tell you which.

If the flue is located in a wall or outside a wall its foundation will be as good as your house's. Old New England homes with central chimneys have a large stone foundation built up in the cellar, and these seldom have more than a stone or two out of place. Most modern flues rest on a sound concrete or crushed rock foundation. But I've seen some middle-aged houses where fireplaces and flues were built on huge planks strung between a pair of stone buttresses in the cellar. Anything like that deserves some more support—a solid foundation or a couple of heavy-duty house jacks put permanently in place.

You may also be surprised to find that your old brick flue doesn't extend all the way down to the cellar. Many older houses—country homes for the most part—have flues beginning on the upper floor or perhaps in the attic. Such flues are supported only by the house framing and were originally con-

Stuffing flue-top with straw-filled sack... ...to find leaks

nected to stoves with runs of stove pipe that were often too long for safety or that ran through sections of floor or lath and plaster wall without proper fire-stop boxing around the through-wall connections. Reactivate one of these with an extra dose of caution.

Now, unless your flue is obviously beyond saving, build a smoky fire in the fireplace or stove, or light an oily rag and put it through a smoke hole. Plug the rooftop opening with an old pillow, a feed bag full of leaves or the like. If there are holes in the flue, the escaping wisps of smoke will tell you where. Mark them for repair. If the flue looks as though it has sprung a thousand leaks, I'd abandon it and build a new one, have it rebuilt, or perhaps keep it for ornamental value. Remember, for each leak you can see, there are probably several hidden behind walls.

If your house is relatively modern and the chimney done by a good mason, there should be no severe problems. If there's a smooth liner in the flue, made up of round or square cylinders molded of fireclay and placed one on top of the other, you've a good, modern chimney. Most likely there will be two or perhaps more flues in the chimney, one for the central-heating furnace and one for each fireplace or stove. About the only problem that crops up in the older versions of these chimneys is leakage from one flue to another. Often the joints between tile liners weren't mortared and in time interflue leaks developed. There is some

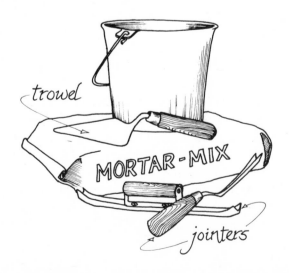

trowel

MORTAR-MIX

jointers

slight chance that sparks could travel from a warm flue into a cold one, drop and start a fire. And the draft of both flues can be affected by the leaks. They can be remedied easily.

# Repairing Faulty Flues

For fixing the outside of a chimney you'll probably need a small mason's trowel, a pointing trowel (like a bent screwdriver) for repairing joints, a bag or two of premixed mortar mix (not sand mix or concrete mix), and a bucket. First, dig out all loose mortar and broken bricks or other building material. Then mix up as much mortar as you think you can apply in a half hour. Wet down the chimney area you will be working on so the cement will mate well. Then use the pointing trowel to fill in joints. Pack mortar in well, but leave a groove in the finished surface of each—protruding cement invites water in. Loose bricks can be mortared back in place unless they have gone soft.

### The Exterior

The portion of chimney exposed to weather may be badly deteriorated, while the inside and part of the outside under cover are in good shape. Here, you can just plaster the entire outside. Wrap tightly with fine-mesh poultry netting, or use metal wall lathing. Just make sure it is wrapped tightly and as close to the brickwork as you can get it. Then plaster with the mortar mix. Do it in two coats, the first to fill in spaces between the wire, the second, before the first has dried completely (12 hours or so later), to cover the wire.

If plastering a chimney, you will want to reflash between first and second layers. Get some copper flashing from a builder's supply outlet if you possibly can; it's expensive, but it will hold up. Lead sheeting is even harder to find but just as good. You may have to settle for galvanized iron or aluminum; the lime in your mortar will react with both kinds of metal, however, and copper is worth the search. You don't need much; a roll of six-inch wide flashing should do. You'll also want a can of roofer's cement or tar and a putty knife. Where the chimney meets the roof, parallel with the eaves, ridgeline and run of shingles (horizontally) you needn't pry up roof covering. Put a thick layer of tar along the roofing at the joint. Then cut a strip of flashing and bend as shown.

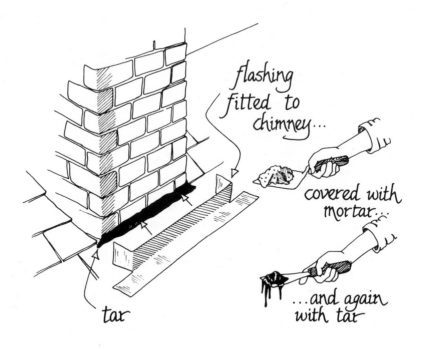

flashing
fitted to
chimney...

covered with
mortar...

...and again
with tar

tar

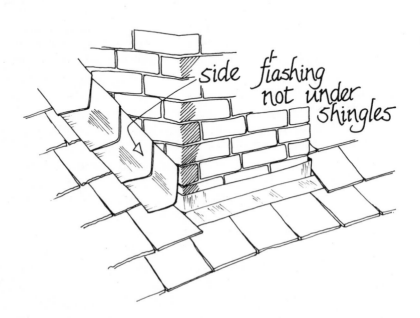

side flashing
not under
shingles

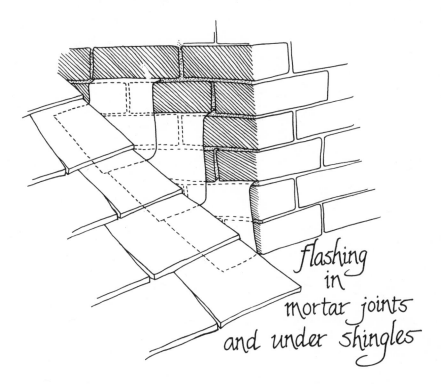

*flashing in mortar joints and under shingles*

Slip bent strips of flashing in under the shingles, starting at the bottom. Press the flashing into the mortar. Now apply the second coat of mortar, covering the flashing on the chimney completely. Once the mortar has dried, put a coating of tar on the horizontal flashing, covering it thickly and completely. If you've not been able to get strips in under shingles on the sloping joints, you should tar them also.

Reflashing a chimney that is not to be completely plastered is about the same job, but you tar along the entire joint in place of concrete—a tar coat, the flashing, then more tar. If the mortar joints are loose enough, you may be able to slip an edge of flashing into the brickwork and cement it in. This makes the best job of all.

All chimneys should have a cap in the form of a low cone with the top cut off to shed water. You can also put on a rain cover, an especially good idea if the inside of the flue is in bad condition. If you live in a particularly windy area, I'd suggest

putting no more than a thin, flat cement cap on at first, running the flue for a while to see how the prevailing winds affect your draft. You may want to incorporate a windbreak into your rain cover. There are also a number of metal and ceramic chimney pot designs that you can shove in or lay on and remove for cleaning. They help alleviate wind problems and also increase flue height, often a quick and easy cure for a smoking fireplace or poor stove draft. But remember: Anything that impedes smoke flow will accumulate creosote. Plan to clean your windbreak several times a year.

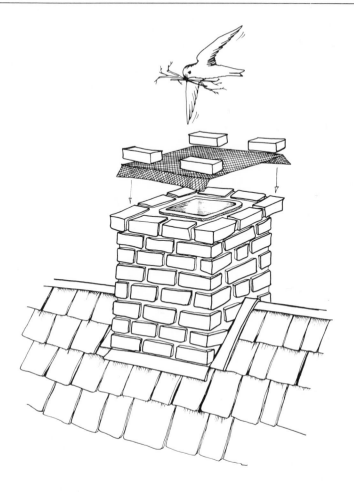

If your chimney is located on an outer wall so one entire side is exposed and in bad shape outside, you can plaster it up just like the top. The worst problem is fastening the chicken wire to the brickwork. I've found that if the brick and mortar is punky enough, sturdy electrician's staples can be driven in between bricks securely enough to fasten yard-wide strips of wire. Fence staples are too soft, as are the staples that come with staple guns, which are also too short. It takes a lot of climbing up scaffolding and much hammering, but a complete plastering job done in your spare time is a lot cheaper than a complete chimney-rebuilding job.

While we're up on the roof, better put some screening over the flue top. It will keep large sparks in and such bothersome chimney dwellers as bats and nest-building chimney swifts out. Check it frequently for creosote buildup and don't expect it to withstand rust-out for more than a year or two.

## The Interior

Reconditioning the inside of flues is chancy. Old chimneys that were used for years of woodburning will likely have a many-inch-thick coating of carbonized smoke deposits as an inner lining. So long as you keep it from catching fire, it is about the best lining you could wish for. But after years of disuse and exposure to rain, this creosote can come loose. Or you may have bad joints and rotten brick inside as well as out. It is a good idea to scour the inside of any old flue. Get a set of tire chains and enough rope to reach from roof to flue bottom. Hang the chains down the flue and pull them up and down repeatedly, from top to bottom and all around. Anything loose will come off in time. You may have a few hundred pounds of junk in the bottom of the flue, but you'll know where you stand. Inspection of the inside is

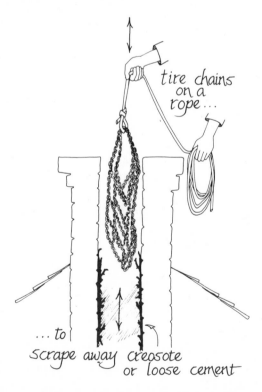

tire chains on a rope...

... to scrape away creosote or loose cement

done most easily at night. Tie a strong lantern to a cord and lower down the flue. If the light reveals that any big chunks of masonry are missing, plan to abandon the chimney or have a mason either rebuild it or line it with fireclay flue liners. If there's only minor damage, there are several things you can do yourself.

*Recementing the Flue*

If all looks sound, go down and light your fire. However, if there are loose joints, broken bricks (or uncemented joints between sections of liner in a modern flue), you have a messy job ahead. First get a bag of fireclay cement or furnace cement from a heating contractor. This is the same stuff you use to mortar a masonry firebox for a fireplace; heat hardens it up to a rock-like state. Next make a flue traveller. Best is a board an inch smaller all around than the narrowest section of flue. Attach ropes to each corner, as shown, so the plate will pull straight up the flue without tipping. Then enclose the plate in a bag of tough cloth filled with rags, straw or other loose material, again per the drawing.

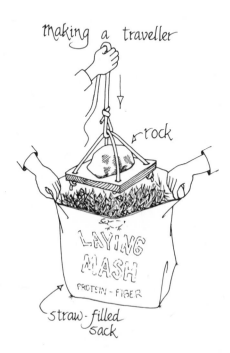

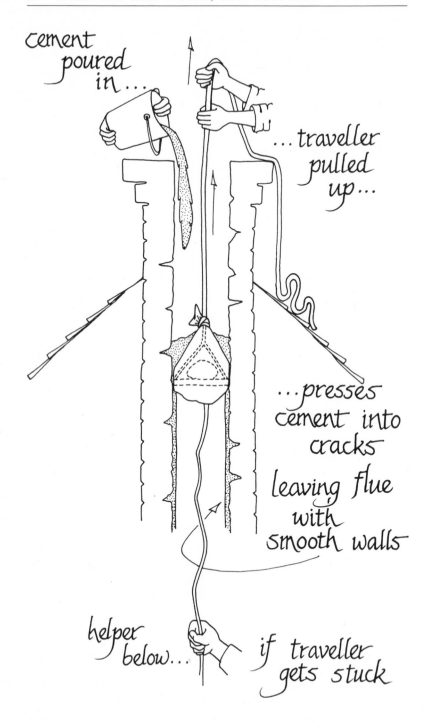

cement
poured
in...

...traveller
pulled
up...

...presses
cement into
cracks

leaving flue
with
smooth walls

helper
below...

if traveller
gets stuck

Now mix up the cement to a consistency of heavy cream. Attach a line to both top and bottom of the traveller, the top line running to rooftop, bottom one hanging down the flue, just in case the traveller sticks and has to be pulled down. What you want is to pull the traveller up to each joint and pour enough cement to fill the cracks down the flue. As I say, it's a messy job. You'll end up with half the cement at flue bottom, and the traveller will likely stick a half-dozen times before you are done. I've done it single-handed and spent half my time running up and down ladders in an impotent fury. One person below, one on the roof and another to mix and haul cement is better. However you do it, the cement is runny enough that it will get into the seams and most of it will stay there. Just go slowly, pulling the traveller up one layer of brick or joint of liner at a time, pour down cement to fill the traveller's moving edges and give the cement plenty of time to move into the cracks. Frustrating perhaps, messy for sure, but a good way to reline an old flue.

Now, here again we must bring up building codes. You may be able to reline your flue so it will give faithful and safe service for years to come. However, even this minor work may negate any "grandfather clause" provision in the code or in the view of your insurance company. As wood heat becomes increasingly popular, the powers that be are going to insist that our flues be "Class A," either ceramic-lined brick or block flues or the insulated-metal type. So, don't waste your time putting in a code-breaker.

### Lining the Flue with Stovepipe

An old-time country practice that sets the citified building code types all a dither is to add a few years to an old flue by running a length of stovepipe up it. I know several people with big old large-diameter flues who have put in Class A insulated-metal pipe to the satisfaction of any inspector. But forget using regular stovepipe unless you are country folks in a live-and-let-live part of the world. I'd get the heaviest pipe available, of stainless steel or chrome plate if at all possible. Then, don't run any type of airtight stove into it. There are more details in the stove chapter, but, in brief, these stoves produce great amounts of creosote that can eat sheet metal away in no time. Pipe liners are for small, hot-burning stoves or freestanding fireplaces and

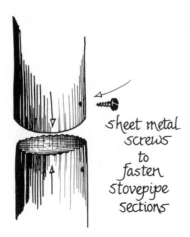

sheet metal
screws
to
fasten
stovepipe
sections

shouldn't be asked to serve for more than a few years while you
save up the dough to put in a permanent flue.

Most any chimney this deteriorated will surely be old and of
sufficient width and of archaic design that you will have no
dampers or odd angles in the flue to complicate the work
overmuch. Simplest is if you have a big old chimney and want to
install a stove small enough that a pipe elbow will go down the
chimney. On the roof, assemble your pipe, elbow on the bottom. With a drill or steel punch, put holes through both pipe
lengths on both sides of each connection. Then fasten together
with sheet metal screws. Needless to say, size of hole should be
a thread's width smaller than the screw.

Now, lower the pipe, fastening as you go till the elbow is opposite the opening in the flue. It will take a bit of doing to attach
a length of pipe to the elbow through a small opening. You may
have to knock a few bricks out. Or perhaps you can cut the pipe
to a one or one-and-a-half-foot length. Then one person can put
an arm through the pipe and on into the flue to grab and hold
the elbow while another works the pipe length on around the
partner's arm. Be careful with this and wear stout gloves. Sheet
metal can cut, especially if you run a hand along a newly sheared
edge.

You can support the pipe at the top by simply running a rod
through holes in each side, the rod ends resting on the flue top.
More complicated, but more permanent would be to provide
full support. At the opening down in the house, by trial and er-

stovepipe
is twisted
and pushed in
around
partner's
arm...

...to
connect
with
elbow in flue

ror, perhaps by snipping ends off a stick, determine the diagonal measurement of the flue. Now get a couple of lengths of angle iron or metal pipe just a bit longer than the two measurements. Use a hammer to wedge the metal into the brick in an X, rods touching in the middle, the X about an inch below the level of the bottom of the hole. Measure the inside flue dimensions and

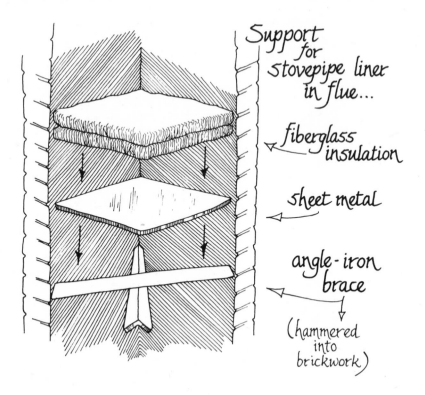

Support for
stovepipe liner
in flue...

fiberglass
insulation

sheet metal

angle-iron
brace

(hammered
into
brickwork)

cut a piece of sheet metal just a bit smaller all around. Roll it up, put it into the flue, unroll it and lay it on the X. Now put several layers of fiberglass insulation on top of this platform. Install the pipe as above.

Back on the roof, wiggle the pipe so it is as near as you can get it to dead center in the flue. Around it pour in just about any loose, lightweight fireproof material you have. Loose rock wool insulation would work if it will fall freely down the flue. I would recommend one of the mineral garden soil conditioners, such as vermiculite. A cubic yard weighs practically nothing, it is cheap in bulk and will not mat down over time. Just test it well to be sure it is fireproof. Some folks have suggested pouring concrete in around the pipe. It would take a lot of mixing and pouring, but it would be permanent and serve long after your pipe is rusted away.

You will want to seal the top of the new flue against water. Put a raincap or downwind elbow at the top of the pipe. Then, with several layers of fine-mesh poultry netting, build a mesh

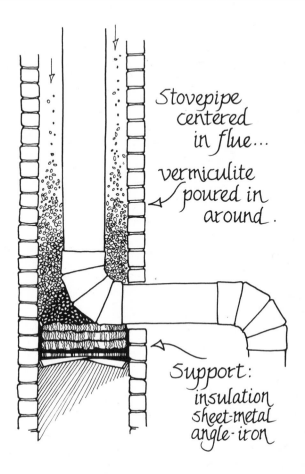

Stovepipe
centered
in flue...
vermiculite
poured in
around.

Support:
insulation
sheet-metal
angle-iron

topping over your inner core. Make it higher by an inch or two at the pipe. Put on enough mesh that it will support your hammer without bending. There should be no holes through it more than an eighth of an inch or so; three to six overlapping layers would do, depending on mesh size. Now, just mix up mortar and plaster the cap so all wire and the top bricks of the flue are covered well by a good half-inch of mortar. It's best to apply in two stages, as with the stuccoing job we did on the outside of a defective flue—the first layer to cover the wire, the finish coat going on about a half-day later. While you're at it, you may want to stucco up the outside of this chimney. If you do, and the old chimney is not about to topple, it should last a good many years.

Lining a flue over an existing fireplace is more of a job. If you plan to use the fireplace, don't even attempt it unless the fire-

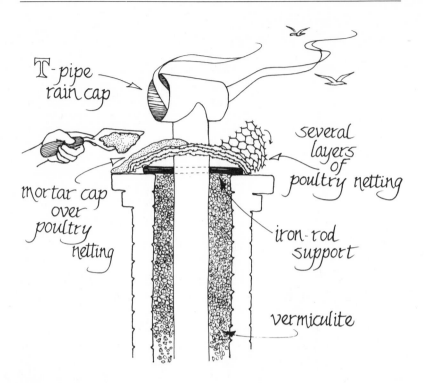

T-pipe rain cap

several layers of poultry netting

mortar cap over poultry netting

iron-rod support

vermiculite

place and smoke chamber above it (see chapter 4) are in serviceable condition and you can get to the bottom of the flue (where it narrows at the top of the smoke box). You will want to put down the biggest diameter pipe you can, and seal the rest of the flue at the pipe bottom. I've seen some situations where the pipe can be lowered into place, then two people with long sticks, one on the roof, one in the fireplace, could ram fiberglass insulation in around the outside of the pipe and pack it well enough at the base to form a decent seal and hold up the vermiculite outer core. The rest of the operation is as described above.

If the fireplace lacks a damper, you can open a hole in the flue big enough to work through and install one (see section on installing stoves). Place the pipe so the damper can be operated through the hole and you've a solution to the problem of installing a damper in a very old fireplace.

It would likely be on the second floor, thus a minor inconvenience, but this is the quickest and cheapest way to control draft in an old flue that you have to reline anyway.

## A Safe Flue Connector

We mentioned earlier that you may find old smoke holes punched in your flue—or perhaps you will be putting in new ones. The problems and solutions are the same either way. The actual job of cutting into a flue and installing a metal thimble to line the hole, permitting a good pipe fit and ease of installation and removal, is straightforward and is covered a bit further along. But the flue connector—whatever lies between the inner face of your room wall and the inside of the flue itself—can cause real headaches.

If you are lucky, the brick of the chimney will be flush with the room wall, perhaps smoothed out with a coating of mortar or plaster with wallpaper or paint over that. If an existing smoke hole is at least 18 inches down from the ceiling and the plaster and room surface around it are rock-hard, all you need do is put in a thimble if one is lacking. If the hole is less than 18 inches from a wood lath and plaster or other combustible ceiling, plug it and hack out another a safe distance down. If the wall is papered, remove the paper in a circle 18 inches in radius around the hole. If you find typical old plaster, crumbly and packed with horsehair or other flammable, fibrous binder, chip it out and replace with premixed mortar/sand mix. You'll have to wash the brick well and keep it wet as the mortar goes on. Perhaps it will be necessary to attach a support of fine chicken wire to the chimney with staples. In any event, be sure the wall around the smoke pipe is flameproof.

Most commonly you will find the chimney has been hidden behind a complete lath and plaster job. Often the flat, thin wooden lathing is fixed to vertical furring strips laid up parallel to the chimney. This whole flammable mess should be ripped out. Best is to remove it at both sides out to the studs—large vertical wooden members that are 2x4 stock in new construction, you name it in old houses—that actually frame the wall. Then put in horizontal framing top and bottom to make a box. If the opening is small and shallow, you may want to mortar it all in. If more than you care to mortar, install the thimble, pack the rest of the opening in the wall with fiberglass and cover all but the thimble with a sheet of fire-stop gypsum wallboard.

Worst is when you find a long smoke connector, such as the

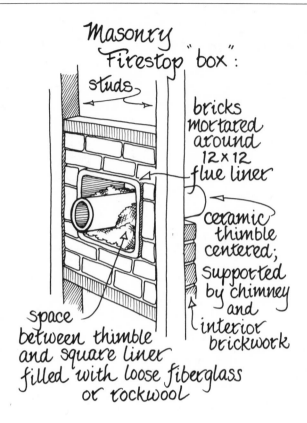

**Masonry "Firestop box":**

studs

bricks mortared around 12×12 flue liner

ceramic thimble centered; supported by chimney and interior brickwork

space between thimble and square liner filled with loose fiberglass or rockwool

amateur-built job we found connecting our woodburning range to the flue. The old brickwork was leaking and the smoke pipe had blistered paint because it was too close to the wall. Your problem—if you have one—will be different, probably unique in all the world. In essence, though, if you find a space of more than a few inches between the flue and inner face of the wall, you will have to build a fireproof smoke connector if you plan to plaster it up out of sight forever. (If you leave a gap in the wall and surrounding partition that's large enough, you can just run in stovepipe. But anything to be hidden away must be fireproof and permanent.)

As with all masonry, I recommend that you hire a competent mason to do the work. He'll have the tools and materials, knowledge of the local fire laws and building codes, and can probably do the work in an hour or two. That's what I say, then I go ahead and do most things myself. Basically, here's what you

## Sideview: Masonry Firestop-box

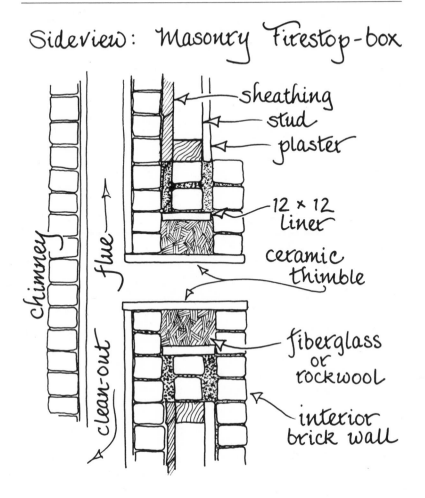

sheathing
stud
plaster

12 × 12 liner

ceramic thimble

fiberglass or rockwool

interior brick wall

chimney

flue

clean-out

want in a flue connector. First and foremost, there should be a smooth fireproof smoke tube extending from the inner face of the flue to the wall surface. Your usual sheet metal thimble won't be large enough, so you'll probably want to mortar in a vitrified clay thimble. This ceramic thimble is really a round flue liner. It comes in two-foot sections but can be cut to any length and is available to fit all standard stovepipes. Any mason's supply house should stock them. This should be surrounded on all sides by eight inches of brick, but you may have to or want to improvise. Here again, an experienced mason is a good man to have around.

A ceramic thimble surrounded by solid brick and mortar is undoubtedly the safest and longest-lived flue connector. How-

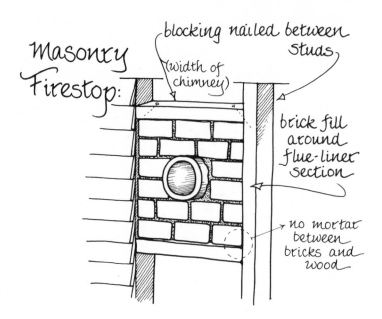

Masonry Firestop:

blocking nailed between studs

(width of chimney)

brick fill around flue-liner section

no mortar between bricks and wood

ever, for any number of reasons you may not want to go into the mason's trade. If so, you can build a firestop box and multiple-pipe, through-wall connector as detailed in the section on temporary-use flues. Or, if you are installing an insulated-metal flue, you can use the fire-stop unit designed for use with the pipe. Just remember: You must have a single smooth pipe from inside the chimney to the inner room wall. Otherwise, sparks might get out into the walls and it's so long, house. Or pyroligneous acid can seep out to stain walls, build up inside partitions or work its way down between the flue and brick or mortar you've plastered on. All unlooked for events.

## Building Permanent Chimneys

There are only two kinds of long-life flues I'd try putting up myself, and I'll say again that you ought to hire professionals to do yours—a prefab metal one and a cement block one. Laying a brick or cement block chimney is a task for a mason, one with a lot of experience. Oh, anyone with a fair eye and a mason's level can raise a chimney that will hold up; you see plenty of home-builts out in the country, and I've built a few.

But anywhere that appearance counts, the unevenness that an amateur is bound to build into the courses will produce an

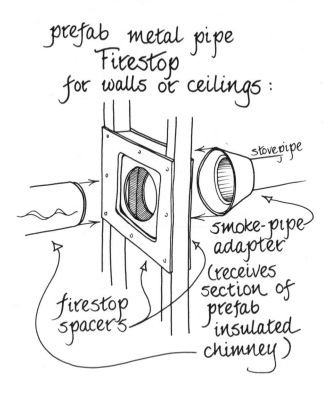

prefab metal pipe
Firestop
for walls or ceilings :

stovepipe

smoke-pipe
adapter
(receives
section of
prefab
insulated
chimney)

firestop
spacers

eyesore; it's a fact that bricks and blocks are so shaped that they have to be laid plumb and true overall course after course to look good. Then too, there's the scaffolding to put up and all that masonry and mortar to lug up the ladder. Every mason comes accompanied by one or two really strong assistants. Indeed, I'd not recommend that most people undertake any mortared chimney. Blocks or bricks get awfully heavy up at second floor level, and if one falls, it can cause a lot of damage.

**Prefab Metal Flues**

Easiest and quickest to install are the stainless steel flues such as the Metalbestos brand available piecemeal at most stove stores these days, in complete kits sold by such large catalog retailers as Sears or Wards and by some—but not all—heat-- ing contractors. As the illustrations indicate, the manufacturer has done most of the work for you. Flue sections consist of two or three concentric stainless steel tubes with fireproof insulation packed between them (where it can't get out and pose a health

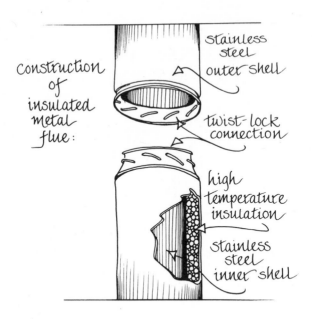

Construction of insulated metal flue:

stainless steel outer shell

twist-lock connection

high temperature insulation

stainless steel inner shell

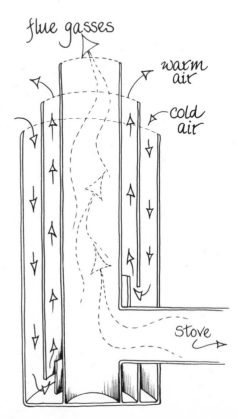

Cutaway of a triple-wall, air-cooled metal flue...

flue gasses

warm air

cold air

Stove

...not recommended for wood heat

problem; this material has been implicated in lung cancer deaths so it's not good stuff to have flying about).

Flue sections screw together easily; the through-wall or ceiling firestop boxes are ready-made as is the flue top and all material you need for going through the roof. All kits and most individual sections of flue come with complete directions. You can get a good set from the manufacturer and help if needed from some stove sellers. But a caution here: The same guy who sets himself up as a wood heat expert by painting a sign that says WOOD STOVES and putting it in a storefront may also be an utterly unreliable stove installer. In few places have our busy code inforcers gotten around to testing and licensing prefabricated flue or stove installers, and even an experienced heating contractor may have precious little experience with this new concept in home heating.

Prefab insulated chimney installation...

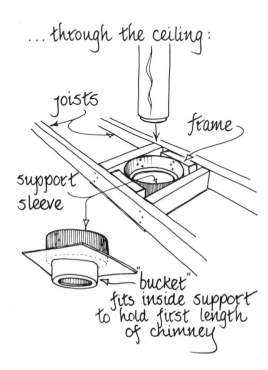

... through the ceiling:

joists

frame

support sleeve

"bucket" fits inside support to hold first length of chimney

So before you hand your bucks and safety over to anyone, double-check his experience. It's best to talk with owners of systems he's installed, and to look the work over yourself. Basically, there should be at least two inches of clearance between double-wall, asbestos-insulated pipe and anything flammable. (There is a three-wall pipe on the market that features an air-cooling system instead of insulation. The pipes are huge and I don't believe the claims made that this—or any other insulated pipe—can be installed with zero clearance from flammable materials. These claims are also made for many brands of insulated fireplaces and stoves and I don't hold with them either. The more clearance you can build in the better. I'd further recommend that you avoid the air-cooled pipe altogether, since the cooling of stack gases is the catalyst for creosote production.)

Be sure your installer (you, if that's who is doing it) knows the difference between stainless insulated pipe intended for gas

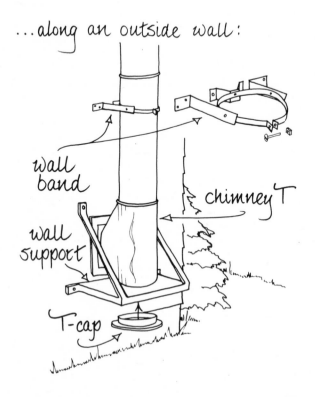

...along an outside wall:

wall band

wall support

chimney T

T-cap

or oil water or space heaters and the much more sturdy, high-temperature resistant "all-fuel" pipe. The heavier-built variety is essential to contain the very hot stack gases of a wood fire.

Like any flue, the metal kind should be run as straight up as possible. Many people are forced to build in curves or elbows to fit the flue to the plan of their home. But every bend reduces efficiency and complicates cleaning—may make it impossible without dismantling the metal flue. The elbow also heats up more than straight pipe as fast moving gases are deflected by it, so the clearance around bends should be larger than minimum, and a fiberglass packing around the bend is often needed for additional insulation.

Check the installer's carpentry. The best job will show the most planning; no through-wall boxes will cut through support-

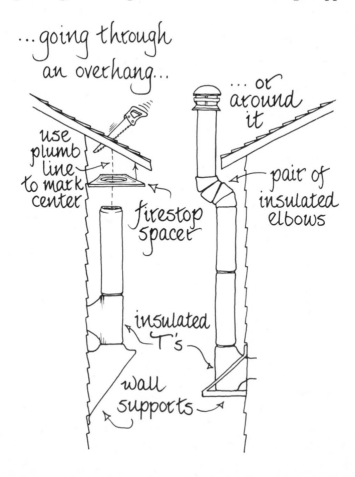

...going through
an overhang...

... or
around
it

use
plumb
line
to mark
center

firestop
spacer

pair of
insulated
elbows

insulated
T's

wall
supports

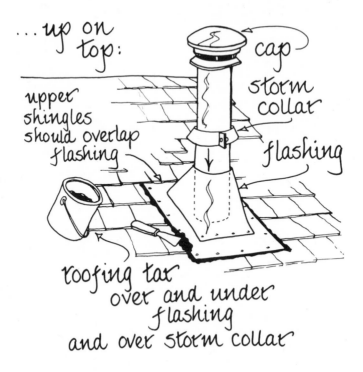

... up on top:

cap

upper shingles should overlap flashing

storm collar

flashing

roofing tar over and under flashing and over storm collar

ing timbers in the house—or if they have, the damage will have been repaired by addition of stout added supports. The roof work should be leakproof, of course. The roof-top opening should be well flashed; there should be a storm collar, a sort of conical fixture around the pipe where it runs up through the roof. As detailed later in building chimneys and ceramic flues, the top should be at least two feet higher than any roof section or whatever that is within ten feet of it and at least three feet higher than the roof at the point where it emerges.

In the house, you'll often see plain stovepipe run right up from stove or fireplace to the insulated through-floor box that is installed in the ceiling. This is widely practiced, satisfies a lot of building codes and is unsafe. The stovepipe can heat up and no length of it should be less than eighteen inches from a flammable surface, including the wood in your ceiling. Liz has drawn in a fireproof sheet on one-inch spacers around the pipe, which will improve safety. Even better in my view would be to hang the flue on the special fitting made for cathedral (sloped, single story) ceilings. A pair of sturdy pins are welded to the sides of a fixture which is suspended from a special hanger. Unlike the standard ceiling support, this gadget doesn't force you to end the flue at the ceiling; you can add more flue, and the best installations I've seen have a safe 18-inch run of flue extending down into the living quarters. Perhaps not too glamorous, but safe. And if you are burning an airtight stove, you would do well to run it all the way to the stove; that way the insulation will keep the flue warm and drawing and will reduce creosote-precipitating cooling of the stovepipe.

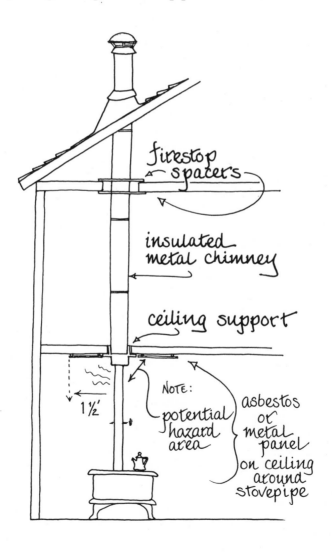

Now, if you decide to install the flue yourself, avoid making the mistakes you are looking for while inspecting another's installation. Follow the manufacturer's directions, check and follow that blasted building code. Oh yes, before even planning your metal flue location, be sure you are familiar with the stove clearances, hearth requirements, et cetera, given in the stove chapter. In essence, plot out where you will be putting stove or fireplace, decide on clearances needed, what fireproof hearths or walls you may need—then start worrying about putting in the flue. You will likely find that the combination of safety requirements, your decorating ideas and the construction details of the house will present you with a complex problem and that a lot of compromises may have to be made—having the stove well out away from the wall, for example rather than inconspicuously off in a corner. Remember, though, safety is the first consideration.

**Masonry Chimneys**

The metal prefabs are easiest for a home handyman to install, and I'll repeat that I'd bring in a mason to build any ceramic chimney. The photo series shows a pro putting up a good Class A, vermiculite-insulated chimney. However, if you must, here's how to build a sound chimney of cement block.

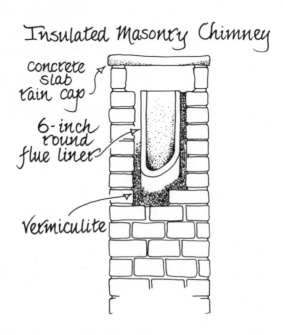

Insulated Masonry Chimney

concrete
slab
rain cap

6-inch
round
flue liner

Vermiculite

Here's how one Class A, tile-lined brick chimney was constructed. Every mason will have his own preferences, and every situation will be different, but basics are going to hold. If nothing else, this photo series should help you decide whether or not you are handy enough with bricks and mortar to tackle a job like this. (I'm not, as I've said elsewhere, and unless you're a mason, you probably aren't either.) The first step is to establish the location of the flue connector, in this case by laying the actual hearth and positioning the stove on it (a wise idea). With the center of the connector pinpointed, the area to be cut out can be determined. In this frame house, the wallpaper, plaster, lath and insulation were removed, the exterior siding and sheathing cut away, any intervening framing removed and the top and the bottom of the firestop hole blocked off with 2x4s. A plumb line dropped from the hole down the outside of the house established a point from which to lay out the chimney's foundation. The foundation in this case was made by digging a hole to well below the frostline and filling it to the top with concrete. Then the mason started laying bricks. At the bottom of the column, a cleanout door was mortared in.

Every few courses, corrugated wall ties were nailed to the house and bent so that their free ends extend well into the mortar. And every few courses, the mason stopped and pointed the mortar joints. At the smokehole, the fireclay thimble was bricked into place. On this job, the mason chose to mortar a short section of 12-by-12 flue liner into the smokehole and extend the thimble through it, insulating the gap between the fireclay units with fiberglass. (A schematic of this setup is found under "A Safe Flue Connector.") Because mortar won't stick to wood, no mortar was used between the bricks and the framing of the firestop hole. Any gaps between the firestop masonry and the chimney column must be filled with mortar. The flue lining begins at this point. Every mason has his own technique for supporting the first liner that's mortared in place and for puncturing it to mate with the thimble. In any case, the thimble should not extend into the flue area and the support setup should not obstruct the passage of a flue brush.

Six-inch round flue liners were used in this installation, and the gap between the liner and the brickwork was filled with a fireproof masonry filler—vermiculite and perlite are two substances used—that also serves to insulate the flue to some degree. As the chimney rises farther and farther from the ground, scaffolding comes into use. Mortar, bricks and flue liners must be put in a bucket and hauled to the working level with a rope and pulley. Here, two floors up, rocking easily in the breeze, is where the masons are separated from the amateurs (like me) with a fear of heights. Just before cutting into the eaves, the void between bricks and liners was filled (as it had been every six feet or so to this point) so flammable sawdust wouldn't get into the space. A hole was ripped in the bottom of the bag and the vermiculite just dribbled in. Using a four- or five-foot mason's level and a pencil, the lines of the chimney were extended onto the woodwork, and the sawing began.

Then the brickwork was laid up through the eaves and flashing installed. With the chimney topped out, the last of the filler was poured in, then sealed in with a good layer of mortar. Brick uprights were set in place and the cap carefully set on top. The cap may be purchased or made by pouring concrete (not mortar) into a form that's easily made from a piece of plywood and some 2x4s. Some sort of reinforcing, like a square of 1x2 or 2x4 mesh fencing just a bit smaller than the cap, is advisable. Pour a layer of concrete, set the reinforcing in place and finish the pour. The cap is made several days before it's needed so the concrete has time to cure. I don't advise mortaring the cap in place, as was done here, because it will make cleaning next to impossible. Heft such a cap and decide whether or not the wind will blow it off. Weight it with loose brick if it seems too light. Inside the house, the wall behind the lovely Morso stove was bricked up well beyond the stovepipe.

The construction of the chimney shown took two people four full, long days to complete, and it took one of them a full day to excavate and pour the foundation, a job done several weeks prior to the erection of the chimney. After a week or two to dry out, the chimney was ready for use.

You'll need a lot of mortar mix—a combination either of portland cement, hydrated lime and sand or of masonry cement and sand—a wheelbarrow to mix it in, a hoe to mix it with, and water, of course (which will freeze before it sets in cold weather, so don't wait till late fall to begin building).

The chimney blocks are hollow and come in a variety of sizes, including solid ones that you'll need for the first few courses. Curiously enough, these blocks—all blocks for that matter—have a top and a bottom; the internal webbing tapers, and the top has a wider surface—to better accept the mortar— than the bottom. Be sure you lay them right side up.

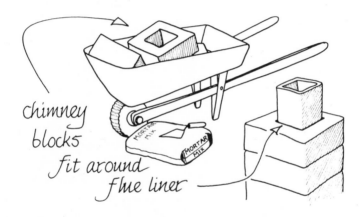

chimney blocks fit around flue liner

Flue liners come in a variety of sizes and shapes, and always in two-foot lengths. Probably the most common liner size is the eight-by-eight, which will serve almost any central heating system or woodburning stove. The eight-by-eight is usually available with a six-inch round hole in one side for the flue connector, so the mason—you in this case—doesn't bust up a half-dozen liners trying to get the hole cut in. But liners run all the way up to mammoth twenty-by-twenties that weigh close to a hundred pounds apiece. The best liners are round ones, but not all masonry suppliers stock them, partly because they present storage problems: the square and rectangular liners stack nicely, the rounds have to be wedged so they don't roll. Most masons are acclimated to using the square and rectangular liners too. The rounds are best because they create the least drag on the stack gases spiraling up the flue.

Scaffolding you can rent from some paint stores or any equipment rental agency, ladders too. Get a pair of steel-toed boots in case you drop a block, gloves, a sturdy spade to dig with, a mason's trowel and a bucket for the mortar. If you have to go up a wood-sided wall or any other flammable wall, get some heavy galvanized wire, wall ties or punched strap iron to make keepers with. Though codes may not so require, I'd build the chimney out two inches from the wood, and with a frame house would use some sort of metal attachment between chimney and house wall every few feet. Next, get a mason's chisel, a small high-temper chisel with an inch-wide point, and a full set of carpenter's tools, if, as is usual, you will be carrying the flue up through the eaves of your roof. Finally, double check your life insurance coverage and double it if the house is over one story high and you aren't used to working at considerable heights. At this point I might add again that a mason could put up the chimney in two days for not much of a fee.

Run your stovepipe out through the house wall. In frame houses, you can use the same fire-stop box installation covered in detail in the final section of this chapter, including cutting out the wall, using triple-insulated pipe and an insulated, galvanized metal box. The inner shell of the triple pipe should be a ceramic or cast-iron thimble long enough to run from inside the room out to the liner of the flue; this will vary from house to house. Or, as illustrated, you can fill the through-wall space with bricks or blocks.

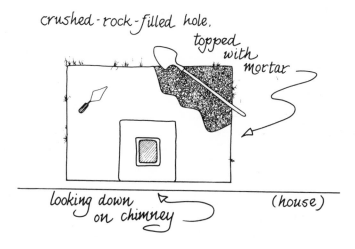

crushed-rock-filled hole, topped with mortar

looking down on chimney        (house)

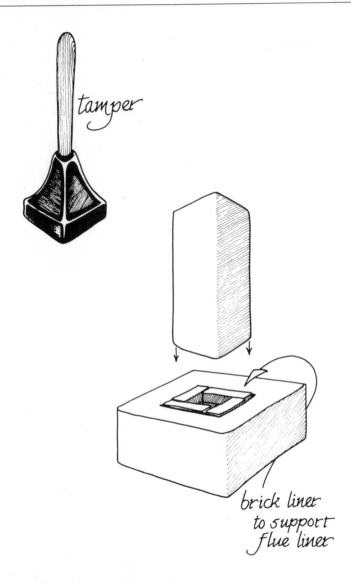

tamper

brick liner
to support
flue liner

Now dig your footing to a foot below frost line or more and three times the width and twice the depth, front to back, of the outside dimension of the chimney blocks. Fill the hole with concrete, or pour in two-inch layers of crushed rock and tamp them down till they are perfectly firm. A big sledge is a good tamper, but you can buy or rent a regular tamper, a heavy, square iron head on a wooden handle. Add rock two inches at a time, tamping well till you have a bed at least a foot thick. Now, pour in a good bed of mortar and lay in your first block, a solid one, center plumb below the center of the smoke hole. Butter on a half-inch of mortar and lay on the next, and on up to a couple of feet below the pipe. Each block must be perfectly plumb, square and seated over the other below. Continue to use a half-inch of mortar.

Two feet below the smoke hole build in your cleanout. Use a chisel to score all around one side of the hole in a hollow block, knock out the middle and put the block on with the opening aiming away from the prevailing wind. In the hole you can put a temporary closure of bricks you will have to dig out when clean-out time comes along, or you can mortar in a clean-out door, available at mason's supply houses. If you can get them, the blocks for the cleanout should have an interior opening a few inches smaller all around than the upper blocks in which the liner will go. Put one more small-holed block over the cleanout and you have a solid base for the liner to rest on. If you can't get such blocks, you'll have to build up the liner support with a layer of bricks, mortared inside the hollow blocks.

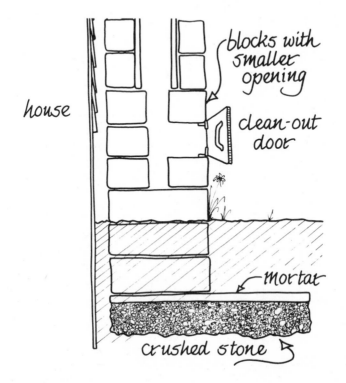

Use fireclay cement to bind the liners. Butter the base block or bricks with it, set in a section of liner, making sure it is plumb and square. Scrape out the excess from the inside. Put blocks on over and around the liner, keep adding liner first, then blocks. Don't put mortar in the space between liner and blocks. When you come to the smoke hole, use the chisels to chip properly sized holes in blocks and liner, and cement in the thimble. Mortar in keepers every six to ten feet.

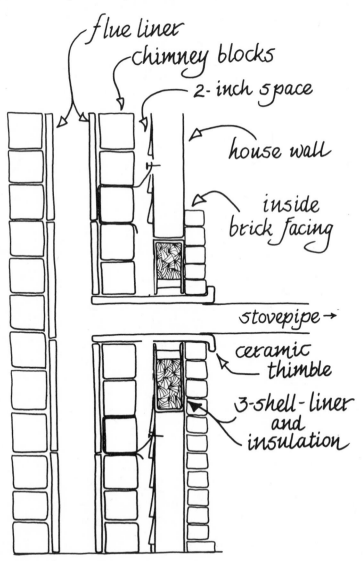

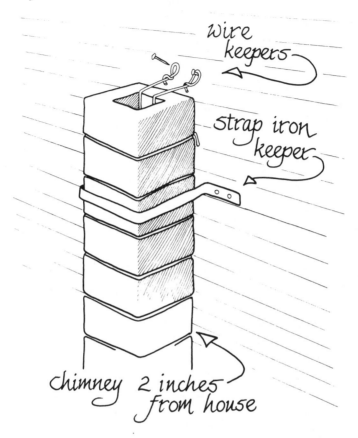

wire keepers

strap iron keeper

chimney 2 inches from house

When you get to the roof, you're on your own. There's all sorts of stuff up there you'll have to saw into: rafters and fascia and scotia and storm gutters and roof sheathing and shingles. Good luck on any sort of permanent residence-type house; I'd call in a carpenter. Just remember to retain that two-inch clearance between flammable material and the flue. Bind in flashing as described earlier and build the chimney to the proper height. Cap it well and put on bird screens, wind deflectors or whatever else is needed. If the space between house and flue bothers you, nail on lengths of overlapping flashing along each side. Let it all cure for a week or more, then light up.

## Flues for Temporary or Intermittent Use

You may encounter flues that consist of nothing but stove-pipe, or in rare cases, old "cattied" chimneys of mud and sticks.

Neither are suited for long and constant use, but do fine where heat is needed infrequently enough that a major investment in masonry or prefab chimneys is not justified. We'll build one of each from scratch, which is pretty much what you will have to do if an existing one is in poor repair or improperly installed (usually the case, especially with stovepipe flues).

### Stovepipe Flues

All the books and reference guides tell you never to build your flue of common stovepipe and few, if any, building codes in populated areas where building codes serve a needful purpose will permit it. And I go along if you're contemplating anything approaching a permanent installation. Stovepipe is flimsy stuff and liable to all sorts of problems under sustained use. But for your two-room log cabin out in the deer woods, which receives no more than a few weeks cold-weather use each season, or your shop in the garage that's heated only now and then, a stovepipe flue is just fine—as long as you take proper precautions.

First, plan at the outset to dismantle and clean out a stovepipe flue after every period of use. That way you'll automatically avoid buildup of potentially flammable soot and creosote. Besides, if you leave the pipe up during several months' disuse, the weather will rust it out faster than a continuous fire. If you'll be using the flue frequently enough that dismantling between uses is too much bother, you will probably be using the place and the stove too frequently to make a stovepipe flue both safe and practical. Build a permanent chimney.

If you can get by with a quick and easy intermittent flue, however, stovepipe can serve. There are two steps in putting one up. (We'll get into installing stoves and building fireplaces later on.) First, you must guide the pipe safely out of the house, through wall or roof. And second, the pipe must be supported and guyed well enough that it will stay up no matter what the weather. If any significant amount of pipe is retained inside the house it must also be secured well enough that it will not fly apart in event of a flue fire.

In the old days, pipe flues for modern oil-fueled space heaters used to be run out through a simple sheet metal plate with a hole in it that was nailed over a foot-square hole in the building. Plenty of old places still come equipped this way.

However, such an installation is dangerous with a wood stove. Unlike kerosene, oil or gas space heaters, fireplaces and wood stoves release great quantities of heat into the flue. And unless you run the pipe inside for many feet—allowing the hot gases to radiate a lot of the heat before exiting from the room—the sheet metal plate can heat up enough to kindle the adjoining woodwork wall, lath or wallpaper.

The pipe should exit a frame dwelling through a fire-stop box we mentioned in the cement-block chimney section. You can purchase complete units from heating contractors who sell the metal, insulated prefabricated chimneys. But for a cheap, easily transported installation for a remote or seldom used location, you can build your own. You'll need materials for a box in each wall or floor or roof you plan to go through. Most folks take the easiest route out through a sidewall or a window. Just be sure to follow the basic principles, which the building industry has incorporated into the prefab fire-stop boxes. Basically, you want the hot pipe surrounded by two circles of dead air, so you make a "three-wall" pipe that will be cool to the touch no matter how hot the inner pipe gets.

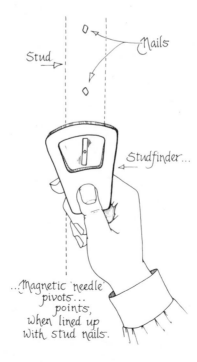

Nails

Stud

Studfinder...

...Magnetic needle pivots...
points,
when lined up
with stud nails.

*Materials for the Stovepipe Flue*

First, you'll need two squares of galvanized sheet metal obtainable from any heating contractor. Get the thickest they have for durability's sake, so long as it is not too thick to be cut easily with tin snips. The sheets should be three or four inches wider than the distance between supporting timbers in your building (called joists in a floor, studs in a wall, rafters in the roof). You can determine this in plastered houses with a stud-finder, a cheap little magnet obtainable from any hardware store. It indicates location of the nails holding floor or wall covering on the framing members. Most frame structures have studs 16 or so inches apart, so the sheets should be 20 inches square. You may find it easier to put the pipe through an existing opening, the top sash of a window, for example. Just get enough sheet metal to cover the opening with enough overlap all around to fasten it to the frame. In addition you'll need a good set of tin snips, hammer and nails, saws and chisels to cut through the walls, two lengths of lumber the same size as the framing (usually 2x4) and the correct size to fit horizontally between two studs plus some sheet metal pipe and accessories you'll find at most hardware stores.

Your stove will have a smoke outlet sized to fit one of the standard pipe diameters: six, seven, or eight inches most likely. For the box get one length of the stove's size and each of the two next larger sizes. For the usual small stove with a six-inch outlet, that would be a six, seven and eight. Pipe comes in two-foot lengths and is usually sold unfastened (open along the seam). This way sections can be nested together and are easy to transport. You'll also want a small bag of loose fireproof material, such as we used in lining the old flue with stovepipe. And finally, get two collars—donut-shaped fittings in the stove's pipe size. A storm collar would be best for outside and decorative collar for inside, but two of one or the other will suffice.

*Construction*

At the site, cut your way through the wall, ceiling or whatever. In brick or stone construction, the opening need be no larger than the stovepipe. Just be sure that no margin of the opening is closer than 18 inches from any woodwork. And, if in going through stone or brick you find a wood inner structure, make a hole as for frame construction. Again keep that 18-inch clearance away from anything combustible.

In a frame building, do your best to cut a square hole 36 inches on a side (or a round one with a radius of 18 inches) through the ceiling, roof or wall. No need to try to cut through the main framing members, unless they are so close together that you won't have a good two inches of space on each side of the multiple-wall pipe we're going to install. Most modern buildings are made with the vertical studs spaced 16 inches from

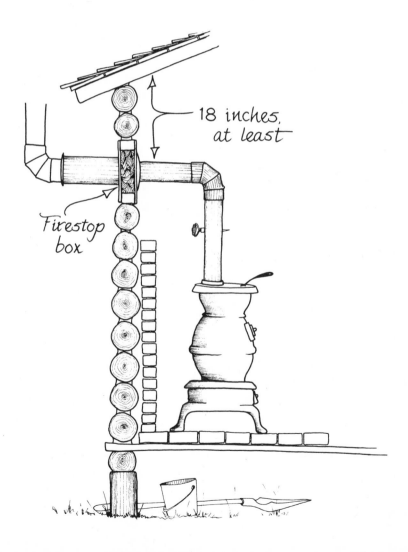

center of one to the other. Most old houses, camps, shacks, barns, and places where you'll be using this sort of flue, were put together the way the builder felt like doing it at the time. Log cabins don't have any studding. Last place I visited with a stovepipe flue was an old steel Toonerville trolley hauled to the edge of a lake for a fishing cabin. Good fishing, too. What I'm trying to say is that we can't give full instructions on how to make your opening. Just do it as best you can, and make it as close to the good and safe 18-inch clearance as you can. Drill yourself a hole in as many thicknesses of partitions as you have to go through, and hacksaw metal, use a keyhole saw on wood, chisel on stucco or interior plaster and your chain saw (pushing in gingerly with the end to get started) in log construction. The hole should be directly above or behind the stove location if at all possible (for details of stove location, see chapter 3). But if for some reason you must angle the pipe, try to have no more than a 45 degree slant. If you want to vent out horizontally behind the stove, the center of the hole should be right in back of the stove's smoke outlet (height of center of smoke hole from floor plus height of whatever hearth you are using—see section on

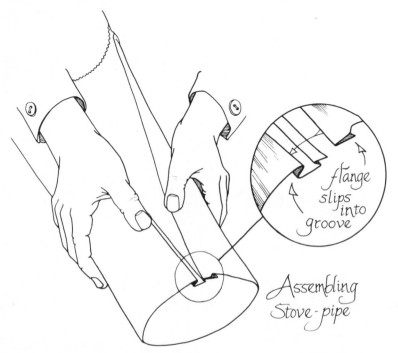

flange slips into groove

Assembling Stove-pipe

stove installation once again). It is better economy to keep as much heat-radiating pipe as safe and reasonable inside the dwelling, so exit the pipe near the ceiling. Just be sure it comes no closer than that 18 inches to any combustible material, and that includes the rafters of your ceiling.

### The Fire-Stop Box

Now assemble the three-part fire-stop pipe. You'll notice that one edge of the seam has a groove stamped in it, the other a flange. Fit them together at one end and work up along the joint, pushing down as you go. It's easy to do this way. If you tried to mate the edges along the entire length in one operation you'd go nuts trying. Put the middle pipe into the outer one and stuff a little insulation in one end good and tight, so the inner pipe is centered. Then set the unit on end and pour in vermiculite or perlite, sealing the space with a little more rockwool or fiberglass. Then do the same with the smallest pipe. Assemble them with crimped ends of all pipes at the same end.

Next, provide a good support on all sides for the metal sheet, which will hold in your fire-stop box. The main type of construc-

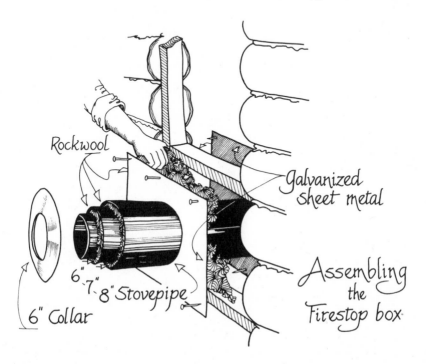

Rockwool

Galvanized
sheet metal

6" Collar

6"-7"-8" Stovepipe

Assembling
the
Firestop box

tion where you'll use this apparatus will be traditional frame. You've got to build a square box, so you must make horizontal members to fit between vertical studding in walls, intrajoist or rafter members in floors or ceilings. It's probably easiest to saw them just a bit larger than the space they are to occupy and hammer them in, to be held by simple tension. You may want to nail them and often there are several inches of lath, plaster and air to go through. Use long nails, and good luck. Or, maybe your walls will be sound enough to hold the (comparatively light) weight on their own.

Now, cut holes to fit the three-part pipe into the two sheets of galvanized; slip the assembled three-part pipe into it, with the crimped ends on the outside. Support the pipe horizontally and pack more insulation into the square hole in the wall, bringing it out even with the inside wall. Then slip the other piece of sheet metal over the pipe and nail it on. The pipe will be two feet long, and you can adjust it any way you want in the wall. For appearance's sake, you may want the inner end flush with the wall so only one size pipe shows in the room. Put a collar over the pipe where it enters the wall and the end of the three-part pipe will not show. And you have a good, safe exit for the stove. Another collar inside will hide the insulation and make for a tidy appearance.

### Outside

There will be a foot or so of three-part pipe showing outside the house. This may be enough to clear the eaves of your roof.

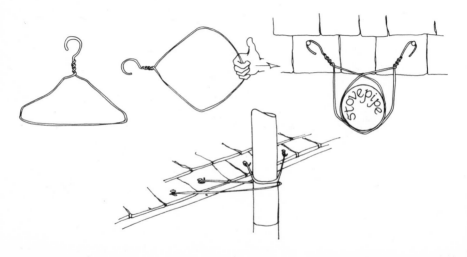

*Cutting Galvanized sheet metal with Tinsnips*

Or you may have to cut a small piece of stovepipe to extend it. There are two ways to support the vertical run of pipe. For short flues typical of one-story cabin installations, just put on an elbow (a four-piece assembly adjustable in angle) and enough pipe to rise at least two feet above the eaves. You can support the elbow with a length of wire running down from the eaves on each side. Another length of stiff wire—a couple of coat hangers—can be bent around the top of the pipe and nailed to the roof or eaves.

Better—safer and sturdier—is to support the flue with a wall bracket and roof-edge metal bracket. You can purchase wall brackets from a heating contractor, or you can make one or have any welding shop put one together for a few dollars. The roof-edge bracket you'll probably make on-site from a length of galvanized sheet metal. If the pipe runs, say four inches from the

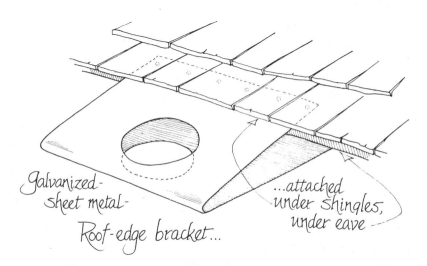

*Galvanized-sheet metal-Roof-edge bracket...*

*...attached under shingles, under eave*

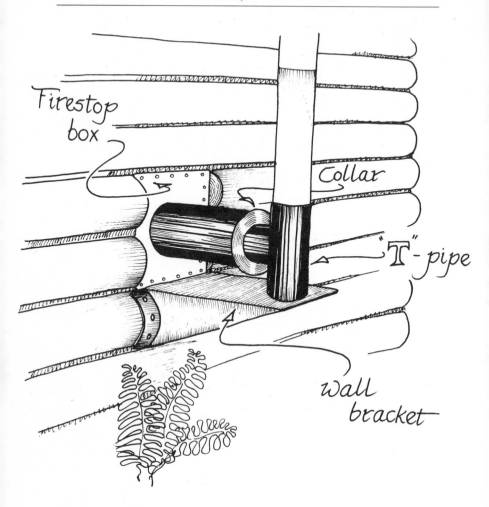

roof, take yard-long pieces of foot-wide sheet metal and bend it in the middle into a V-shape. Cut holes a bit bigger than the pipe near the bend in each half, run the pipe through and nail the ends to roof and eaves.

At the bottom, put a T on the end of the inner pipe of the three-part assembly with the other collar placed to seal the outer end—to keep water and critters out of the stuffing. Attach the shelf bracket so the bottom of the T rests on it, then put your pipe on top of the T. It is a good idea to support the flue at each junction of two pipe lengths, should you have to go up a high wall. Stiff wire looped around the pipe with ends nailed to the wall is fine. At the top put on a rain-cap, still sold in some

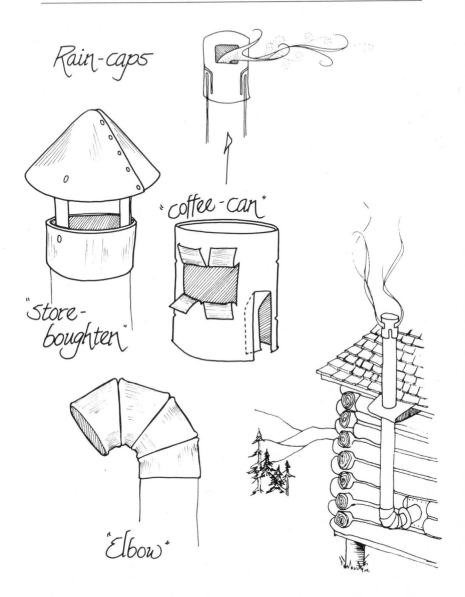

*Rain-caps*

*"coffee-can"*

*"Store-boughten"*

*"Elbow"*

hardware stores. You can make one yourself easily enough, though, from an old can. If there is a lot of wind, another elbow put up top and aiming downwind may be the best cap. Finally, it is a good idea to fill the bottom of the T with sand. This will add a bit of weight to the bottom and increase stability in a wind, but more important will prevent any loss of draft from air getting in between pipe bottom and the shelf.

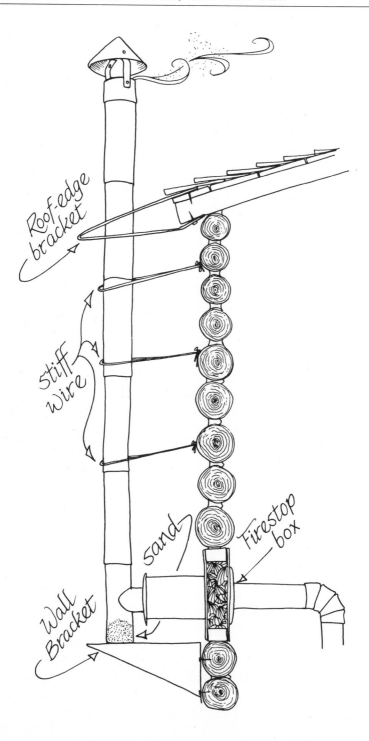

Roof-edge
bracket

Stiff
wire

sand

Firestop
box

Wall
Bracket

Make sure that all supports fit loosely, so disassembly will be easy when it's time to leave. Just pull all pipes apart from the bottom and slip the flue down. (Unless it's very tall, you should be able to assemble it again when the time comes from the ground up.) Shake out all accumulated soot and put the pipe in out-of-the-weather storage. The only chore remaining is to seal up the opening through the wall. Easiest is to buy an end cap. They have handles for easy removal. Grease the end of the pipe, though, or else pipe and cap may rust together. Lacking a cap, a coffee can may serve. Or you can just push the pipe inside and tack another piece of metal over the outside hole.

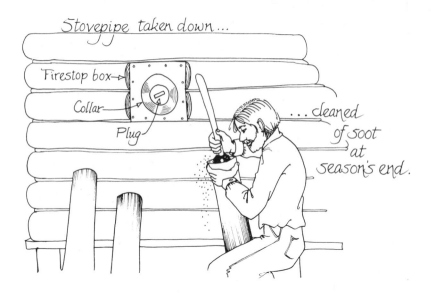

Stovepipe taken down...

Firestop box→

Collar

Plug

...cleaned of soot at season's end.

## Mud-and-Stick "Cattied" Flue

The very first chimneys built by the colonists in America were of mud and sticks, and you can still find a few original cattied flues in the backwoods of some of the southern states. They do require a lot of maintenance, and they can catch fire if neglected or overheated. Together with the wood shingle or straw-thatch roofs common during the first years of the colonial period, these cattied flues were responsible for so many fires that they were outlawed in many cities. But they are fun to build, will last for a good while and serve safely if kept up. And

the "cattied" mud and stick chimney

they cost nothing, require you to pack no more into the cabin site but your own muscles, the food to fuel them and a piece of string.

First job is to find a deposit of the proper mud. A good forest loam won't do. Neither will a sandy soil. What you want is clay, and the finer, gummier and stickier the better. (Some of the best cattied chimneys are found in the state of Georgia—red clay country.) Dig down to subsoil, scout the creek or riverbeds within easy walking distance from the site. Then make a sausage-like roll of each sample of likely mud, let dry and put in the cooking fire. Leave them there through several good fires. Then test. Any that crumble after baking are no good. The ones that harden up are the better choice, the one that is hardest to break and that makes the most musical noise when you tap it is the best of the lot.

Now, pick your flue site for either a stove flue or a fireplace. (If you're building a fireplace under the flue, follow directions in the fireplace construction section; for a fireplace, I'd try a stone-and-mud Rumford, though I've never made one so I can't tell you how.) Cattied chimneys lack the integral strength of a masonry or stone type. So, you should give them as much support as possible, by running them up the highest wall of the structure they will be heating, the peaked end wall usually. Try to satisfy the basic rule of flue construction and plan to have at least two feet of chimney rising above the nearest line of roof within ten feet laterally, so the roof will not interfere with airflow upward out of the chimney.

### Constructing the Chimney

Up against the cabin wall dig a hole that is a foot larger all around than the flue base is to be, and be sure to dig down to well below frost level in the North, or to good solid bedrock or subsoil in the warmer parts of the country. Fill the hole with

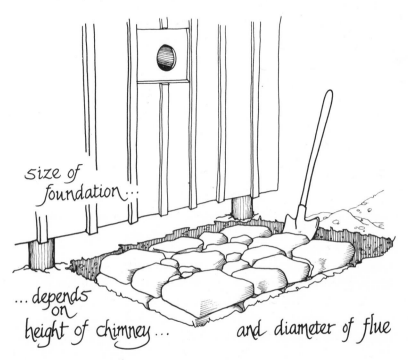

size of
foundation...

... depends
on
height of chimney ...        and diameter of flue

rocks if you have them, putting them down in layers, "cement-ing" each layer with a thick clay and water slurry as you go. If you've the time it is best to let each layer dry for a day or two before adding the next.

Once the base is down, dry and settled, you can begin build-ing the flue. (If you lack rocks for a foundation, begin the flue at the bottom of the hole.) First, decide how high it is to be—the top some two feet above the ridgepole if you are building near the center of an end wall as recommended—and how big around at the top—three times the size of the flue opening. For a small stove needing a six-inch flue, the top should be 18 inches on a side. For an eight-inch flue, the top should be a full two feet on a side. Now for each foot of height you will want to add on three or four inches in base width. If the top is to be 16 feet high, the base would be four or five feet wider than the top. Say, six feet wide for a six-inch flue, seven feet wide for an eight-inch flue. You can add on a foot or two of depth too—extending out from the wall. The added heft at the bottom is to provide stability and a good slope so rain will run off.

Use your string to outline the form of the flue on the build-ing's wall. Tie a rock to one end, tack the other to the roof in the center of the roof opening and score or draw a line along the string. You can also outline the outer sides by tacking the string to the outer limits at top and angling out to the greatest bottom width and marking the line.

If you have stones, it is good to lay a stone base up to fire-box floor or the hole where the stovepipe exits your building. But don't build with stone much above shoulder height unless you have mortar and a conventional flue lining and the need for a permanent, constant-use flue. A mud-and-stone flue could work loose too easily and a wobbly rock pile that high is a real danger to passing deer and other folks.

In stick-and-mud construction, the sticks act mainly as an internal scaffolding to hold the clay in place till it can dry and be baked to hardness by the heat of your fire. You lay the sticks out in an overlapping pattern of squares and pack the mud in around them with your hands, a trowel or small shovel. There are several ways to accomplish this that I know of and I have tried all of them out at one time or another, and you'll probably come up with some ideas of your own. You can use a single column of

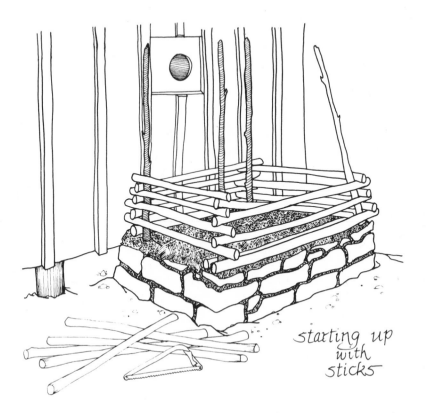

*starting up
with
sticks*

interwoven sticks or use two, one of smaller sticks inside the
outer one. Just be sure you have a good two inches of mud lining
the inside of the flue. If you like, run "corner posts"—straight
saplings with twigs trimmed to an inch or so up the corners of
the flue, just inside the outer sticks. Just make sure all sticks are
green, so they will shrink along with the clay as it and they dry.
They should be perhaps an inch in diameter at most, a half-inch
at minimum. In time they'll turn to char in the chimney.

### Completing the Flue

At most wilderness camps a goodly portion of the food supply usually comes in cans. The only mud-and-stick flue of any ac-
count I ever helped build was for a little 15-gallon gas-can stove
with a four-inch smoke hole. The inner liner of the chimney was
made of fruit juice cans of the just-under-two-quart size. The
flue went up at the rate of juice consumption, one or two cans a
day. The ends were not quite cut out, then bent out and into an

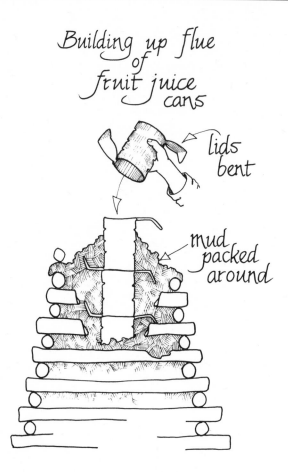

Building up flue
of
fruit juice
cans

lids
bent

mud
packed
around

L-shape, so as to bind in with the mud and sticks. This made for a good, smooth flue lining and reduced need for careful troweling inside the flue; we just added on a can and packed mud around it. With a flattened can on top during disuse to keep out water, the little flue served as long as we needed it and may be standing yet for all I know.

It is a good idea to use rocks, flattened cans or sheet metal to cap the flue. This will keep water damage to a minimum. And plaster as much mud as you have time for on the outside. Be sure to slope the sides well and dig water channels around the base so rain will run away from the flue.

Going in to a stove flue, leave a hole in the flue that is about twice the size of the stovepipe and locate the pipe in the center.

Looking down
into
the Flue...

...after the mud goes on

(A large tin can with both ends out is a good tube through the flue side.) The flue will shrink and if you pack the pipe into the masonry tight at first, it will pull the pipe down, crush it or pull it through the wall. Other accessories you might try include wall ties, to fasten the flue to the building more securely. As you build, mortar in lengths of tin can or stiff wire bent in an L-shape. Have the long part well into the chimney, with the shorter part extending horizontally along on either side of the chimney. After the flue is well-dried and shrunken down, staple the ties to the cabin. Or you can form small circles in the other ends and nail them on.

Burrowing insects—several kinds of small-colony wild bees and wasps in particular—love to dig their spring brood-rearing nests or fall hibernating burrows into nice, warm mud chimneys. In time they'll have your flue spouting smoke from a hundred places. I suppose you could mix chlordane or another bug killer in the outer layer of clay, though the downwash will kill everything in the soil for yards around. An annual spring application of plain whitewash will deter them more safely. Or you can lug in several sacks of premixed portland cement. Mix up a slurry of clay and cement. (I've never bothered to cement up a cattied chimney, so can't suggest a proper ratio, but would start

if it rains
before mud
has
hardened...

chimney should
be covered
with
overlapping
layers
(...sheets,
plastic,
brush...)
and
weighted
down...

lest it all wash away...

out with one-quarter cement and the rest clay) and plaster it on. Be sure to wet the outside of the flue before plastering, though, so the mix will hold.

Fire up a mud-and-stick flue gradually. Build no fires for a week as it dries, then small ones for the next few weeks till all is dried down. If the clay is good quality, you should have few problems with the inner lining, which should bake good and hard. Keep the outside plaster in good repair and the flue should last for a long time.

There are any number of additional embellishments you could add. Build in tubular lengths of chicken wire in place of or in addition to the sticks, and the structure would be a good deal stronger. A clean-out hole for a chimney used with a stove could be included in the outside wall—just build in a juice can about a foot below where the stovepipe enters the flue. When the can rusts out it's time to dig out the accumulated soot and the can. Then plug the hole again by plastering in another can. You can increase the flue's life by wrapping it in weatherproofing, tin-can shingles or strips of building paper.

Still, for an intermittent-use flue, I'll stick with stovepipe, leaving the time-consuming mud-and-stick construction for the really backwoods hunting camp, for recreation or just to prove to the Scout pack that a real working flue can be built with materials that nature offers gratis—just like the wood used for fuel.

## CHAPTER THREE
# Heating with Wood Stoves

As Ben Franklin realized, a stove sitting out in the room, surrounding the entire fire and radiating warmth from front, top, sides and back is the most efficient heating device to put at the bottom of your flue so we'll do that first. If you've enough years on you, you'll recall the stove installation featured on Norman Rockwell's covers on the old Saturday Evening Post. Every odd winter a cover would feature the inside of a New England general store. A big potbellied stove with its nickel-plated footrests and towering smoke pipe occupied the place of honor, smack in the center of the scene. An odd assortment of chairs, a checkerboard on the pickle barrel, a brass cuspidor at the back near the stove for those who couldn't get an angle on the open fire door plus a fair sampling of the retired farmers from the township completed the scenario. But the central figure was that stove, pouring out heat, beckoning companionship, drying gloves and providing a crackling fire to ponder during lulls in the conversation.

Central heating units may be more efficient and bother-free, fireplaces more aesthetically pleasing, but nothing on earth heats better than a wood stove—except perhaps the sun on a fine summer day. The stove fairly radiates good cheer as it warms up your outside, and the hot water in the teakettle on top is right there 24 hours a day, waiting to warm up your insides. If you're just in from a hike in the snow, put your boots right up on the footrest so they almost touch the stove, though they'd best be unlaced first. Wet soles can overheat and stew your toes if you get too engrossed in the checker game.

Of course, now as back then, the potbelly is just one of myriad stove designs developed over the years. Stoves come in all sizes, from a tiny little foot warmer you fill with coals to keep

away the chill during winter meetings to big furnaces that take four-foot-long cordwood and can power a central-heating system. Designs vary too from ornate Victorian parlor stoves to austerely simple Shaker designs. Construction materials include steel, stone, soapstone, brick, and ceramic tiles as well as cast iron. You can get them in plain black or in fireproof enamel in any number of colors. And the variety of stove designs is becoming greater today than any time since the real heyday of stoves, the nineteenth century. They come open so you can see the fire, or closed tight for most efficient combustion (and some have both features). You can find wood heaters with or without cooking tops. A few small European designs can even double as ovens, and then there is the kitchen range, a separate topic altogether.

Why don't we go somewhat into the recent history of stove development, capacity of heating stoves in general, then try to come up with some guidelines to help you select a location, size and design to meet your needs. Then we'll discuss the major designs available, and how to install and operate them (safely above all).

One thing we'd best put right out front, though. There are no absolute truths in wood heat—except for the safety rules, and even there you'll likely need some elbow room. Every wood-chopper has his own grip on the ax, every woodburner has his/her own favorite stove or fireplace, an individual way of firing and cleaning it, a single-minded preference for kind and seasoning of wood, ad infinitum. And isn't it great? The technology is so nontechnological that we can all be our own experts, all just as right as the next ash sifter, even if we differ in our opinions by the breadth of half a woodlot. I'll try to lay out the generalities where I can. But when I express a preference for stoves such as our Jotul #4 Combi, don't you pay any attention. I can show you four dozen people who think I'm nuts; they swear by their Shenandoah or Ashley Circulators, their Lange or Styria airtights, their 1860s era illuminated heaters or their Riteway central furnaces. If there is any group of people who "do their own thing," to use an overworked phrase, it's woodburners.

## History of Wood Stoves in North America

I just implied that woodburning technology is non-technological. But I don't mean that it is simple-minded. Rather, it is a technology that you and I can understand, not all laser beams and computers that only a college professor can figure out. However, back in its heyday in the late 1800s, a simpler time, granted, wood stove design *was* high technology. The production of stoves was a major, highly competitive industry with some of the best thinkers on earth working to make their stoves the most fuel-efficient, clean-burning, durable and attractive to the Victorian eye. As we'll see later, some of the old stoves are masterpieces of ingenuity, containing dozens of iron parts hand cast from handcarved molds, hand fitted and burnished—items of elegant furniture on the outside, fiendishly efficient fuel-misers on the inside.

However, then as now, heating with any solid fuel involves work. And when cheap petroleum came along to save your great granddaddy and mine the drudgery of hauling wood in and ashes out, he happily relegated the heating stoves to the back of the barn. Unfortunately, out with the stoves went all the know-how accumulated over hundreds of years of wood and coal burning. When the energy crunch hit our generation in 1973 there were only a handful of stove foundries in the United States and Canada making what the diminished market demanded, mainly ornamental modern-style Franklin fireplaces for suburban rec- reation rooms. Those few serious woodburners still hanging on in the rural fringes had pretty slim pickings in stoves. When Louise and I went shopping for woodburners in the late 1960s we couldn't find anything that fit our needs and budget, so set- tled for an unmatched set of antique designs.

## The Energy Crisis

Well, with the energy crisis, the demand for wood stoves went through the ceiling. People suddenly were willing to buy anything. One stove seller we know had people asking for old- style chick brooders—anything that would kindle a stick of wood and keep the oil or gas bill down. The suddenness of the demand understandably swamped the small firms that had kept making a few stoves through the petroleum era. The thousands of old stoves mouldering in barns and back lots were either in sad repair (and no one around at first knew how to fix them) or were snatched up and priced all out of sight by antique dealers.

As they say, there isn't such a thing as a vacuum in nature, especially the human-type nature of profit-minded business people. And plenty of folks moved to fill in the shortage of wood stoves. First came the imports. Though solid fuel had been largely forgotten in petroleum-rich America, it wasn't so in Europe, where the only oil of much substance has just recently been found in the North Sea. For much of Great Britain, Ireland and the Continent—the chilly upper reaches in particular— home heat often as not means coal, peat or wood. And the year following the oil embargo, ads appeared for little iron box stoves from Scandinavia that burned wood "like a cigarette." These stoves had a baffle down the middle of the firebox so heat would have to work hard to get out the smoke pipe, losing much of its heat to the room in the process, and a door having an asbestos gasket, tight latch and fit, making it airtight so you could control oxygen supply exactly. People gobbled them up faster than the ships could bring them in.

The next year, native ingenuity not having been put out to pasture with the native iron stoves a few generations ago, ads appeared side-by-side with the import promotions for a growing number of home products—at first all made of welded steel, a few with cast-iron doors or whatever. Scandinavian stoves (and imports from countries farther south) were still coming in, but were scarce, so people gobbled up the steel designs as fast as the shops could turn them out.

Le Petit Godin

## The Problems of Growth

But then, only a few years after the initial oil scare, the bad news began filtering through the growing woodburning community. First was disappointment in the little imports. Many people, as blissfully ignorant of the realities of wood heat as were Louise and I when we first tried it out, had expected the little hundred pound imported stoves to heat an entire house on practically no wood. They found themselves cold and their chimneys getting clogged with built-up creosote. I've read more than one critique that suggests that the early promotion for the imports was misleadingly optimistic. I don't agree; optimistic they were to be sure, but if any misleading was done, it was done by stove-shop owners who didn't know their trade and buyers who knew zilch about wood heat. Well, who did know much before we all had to learn the hard way? There surely are a few sharp operators in the stove business—I know one or two myself. But I'm convinced that most of the problems people have had (and will continue to have for the near future) are simple products of ignorance. We're having to learn wood heat all over again.

The most recent casualties of our facing up to woodheating reality are some—just some—stoves of welded boiler plate. Poor welds are becoming unstuck, doors are falling off, air leaks are developing. Some models with ingenious pipe systems used to circulate room air or water through the firebox are coming into question; one manufacturer has just doubled the pipe thickness in its air circulation system. I for one shudder to anticipate the day the pipe in an early model fails and a great shower of live coals gets blown out into someone's living room. And we're hearing reports that older models of one of the most popular steel stoves with a stair-step kind of top are beginning to warp seriously, ruining their graceful lines and posing a potential threat of breakdown.

And that's just the beginning. In essence, nearly every new wood heating unit developed since the energy crunch hit is an innovative idea put together by an entrepreneurial individual with no more experience in wood heat than the next guy. Or, as one disillusioned stove seller put it, "Hell, everyone with a welding outfit is turning out jerry-built stoves with some sort of harebrained new gimmick. In a few years when the stoves fall apart or burn someone's house down, they'll go out of business and back to hair dressing or wherever they were before jumping on the wood stove bandwagon."

A bit harsh, perhaps, but pretty knowledgeable in my view. I mean, haven't you read articles about the clever backyard welder or bored banker or out-of-work stock broker who came up with this dramatic new stove design and is now making a thousand of them a year? Not to suggest any evil motives. But I do believe that too many people—buyers as well as sellers— have assumed that wood heat is so simple, so non-technical (especially when it comes to making and using wood stoves) that any idea will work.

And, unfortunately, as demand has continued to increase, people all over the world seem to be duplicating the mistakes, perpetuating for the time being the problems that cropped up at the first rekindling of interest in wood heat. I don't know how many copies I've seen of that little Scandinavian box stove; one is being imported from Ireland, several more are being cast in the United States. Not that it's a bad stove design when properly installed and operated in a room small enough for its limited capacity. But I fear that a lot of people will be disappointed, perhaps novice stove buyers, who still expect impossible performance—perhaps stove makers who followed the crowd that followed the first far-fetched expectations for a lovely, but low-powered little stove.

And there are more examples: I can't begin to predict how many more copies we'll see of that stepped-stove design that is reportedly warping or how many other designs will be put together by well-meaning but inexperienced amateurs.

The emphasis now seems to be on innovation, newness in design: a more complicated baffle system, draft control or hot air circulation system. Some stoves of course will prove out well over time. The saddest thing in my view is that too many in the rush for wood stove bucks cooked up their new idea and rushed it to market without the lengthy testing any stove design demands. In effect, we the consumers are being the guinea pigs and a lot of us will regret it. Even sadder is the emphasis on novelty—unproven novelty—in design, when every single "new idea" I've seen come out exists in a hundred different and

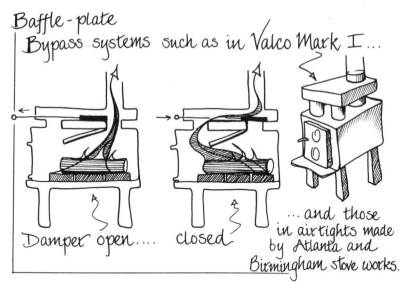

Baffle-plate
Bypass systems such as in Valco Mark I...

Damper open....    closed    ...and those in airtights made by Atlanta and Birmingham stove works.

time-proven forms in the old stoves gathering dust in barns throughout the snow belt.

**The Years Ahead**

Our knowledge of wood stoves is still incomplete, still evolving and probably will always be. However I'm convinced that the worst is behind us, that stove buyers are becoming more knowledgeable, more discriminating, that stove manufacturers aren't far behind and that all else failing, if the public doesn't insist on improved stove manufacturing, installation and operation standards, our friendly government regulators will do it for us. (In woodburning Sweden, with its cradle-to-grave government control of just about everything, there is a semi-annual flue inspection. You have it cleaned or go cold. I for one hope that both suppliers and users of wood heat, especially wood stove business people, adopt enough self-regulation that we don't have chimney inspectors or any other bureaucrats telling us what to do.)

Enough grumbling for now. Be assured that there are good stoves available and that practically anything that can hold a fire—even an old oil drum—can serve as a good and safe wood stove if handled properly. Let's try to help you choose and install the stove or stoves that best suit your needs. First, though, comes some planning: where is the stove going to go?

## Locating the Stove

The first location you might consider is to put your stove in the cellar, if you have one. Or perhaps put one of several there. One notorious disadvantage of heating with stoves and other space heaters is cold feet; without a central furnace or other heater in the basement, your floors are going to be icy cold in winter everywhere on the ground floor but near the stoves. The upper stories will have warm floors. Though we do have a cellar stove, we don't use it to warm the floor constantly, just when we want to work in a warm basement. Sheepskin slippers keep our feet warm, and the children's playroom is on the second story where they can go to sleep on the warm floor if they want without a chill.

### Heating Efficiency

The best first-floor stove location surely must be smack in the center of the most-used room in the house, with a register or grating cut into the ceiling above to let warmth into the second-story room where baby sleeps and grandmother beds down when she brings her easily chilled bones for a visit. With a stove in the room's center, heat will radiate out all around, then rise to

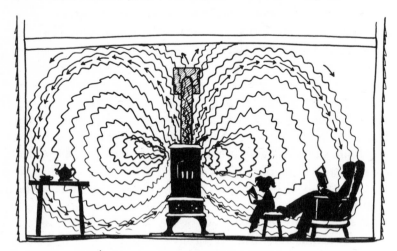

*donut pattern of heat and air flow*

Heat circulation in stove-heated and adjacent rooms.

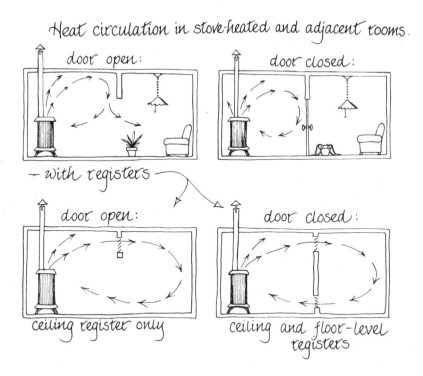

door open:

door closed:

— with registers —

door open:

door closed:

ceiling register only

ceiling and floor-level registers

the ceiling, and flow outward; as it meets the relatively chilly walls, it will fall and flow back along the floor to be warmed up again. The flow is sort of donut-shaped.

Few people will want a stove dominating the center of the living room (though that's where it should go if at all possible). Next best would be on a wall near the center of the house by the stairwell going to the second story if you have one. That way, heat will move throughout the whole house, though the room in which it is located will be the warmest. Until I discovered how much plaster and lathing and woodwork I'd have had to remove for a safe installation, we planned to put our big Combi—which offers an open fire in the evening, but closes up for efficient operation at night—directly into the mid-house flue, even if it did serve the central furnace too. That way the whole house would have been better warmed as the location abuts a big door leading to the stairs. But, we had to settle for another location, and for reasons of economy and ease of installation and maintenance, we decided to vent it into a new cement block flue on an outside north wall.

an airtight stove with doors...

...for use as a fireplace -- and to burn open and hot. Reduces Creosote.

Morso model 1125

Now the Combi heats air that circulates around the lower story, mingling with heat from our woodburning kitchen range. Some gets into the hall and on upstairs, but not much. We cut registers into the bedroom right over the stove; Sam opens it when he gets chilly, closes it when hard play warms him overmuch. There was already a register over the kitchen range; baby Martha's crib went right over it. Now that she's a grown up four-year-old, and can stay well tucked under a set of covers the night through, her bed is moved to the back wall.

### Tying into the Flue

Stove location will also be affected by your home's chimney situation. They can go into existing flues—either through old stove holes, new ones or a fireplace. The illustrations and photos show good examples of each. You may find that a preferred location on an inside wall justifies buying an insulated stovepipe flue. On an outside wall, you can install a prefab or build an outside, ceramic chimney as we did. All the problems of wind direction and velocity brought up earlier will factor into your choice of flue and stove location. Just remember that a stove functions exactly the reverse of most central-heating systems, where radiators or registers are located along the outer walls of the building. There, warmth pretty much circulates up and down the wall, keeping wall and adjacent air warm. Your stove, a single, centralized, radiating heat source, pours its heat out at whatever or whomever is sitting near it, but most of the energy flows straight up, then along the ceiling to a cold wall and down again. There are several two-dollar words that describe these patterns of heat and air flow, but I've never been able to keep them straight. Just remember the donut pattern of air flow, and position your stove so it puts warm air where you want it most. And remember, the air at the extremes of the donut isn't necessarily cold—especially if your stove has enough heating capacity for the rooms. It's *relatively* cool, compared to the 1,000 degree F. plus temperatures inside the stove.

## How Many Stoves?

Now, you must decide how many stoves of what heating capacity you will need to serve the space you want to heat in the position or positions you've selected. There are so many variables involved in the decision that I can't come up with any general rule that makes practical sense. You've got to factor in your flue situation, its size, location inside or outside the house, position of flue and house in relation to wind flow, plus your area's weather, amount of insulation in the house, number of floors needing heat, and on and on. Here's where no book can help you much; you've got to do your own figuring—perhaps with the help of an experienced woodburner or good stove seller.

There have been a few attempts to put definitive numbers on heat outputs of various stoves, unfortunately, all tests having been conducted on the new, well-advertised designs. The best technical information I've seen is in the first half of *The Woodburners' Encyclopedia* by Jay Shelton, PhD. Problem is, from my country boy's perspective at least, you have to be a PhD to understand it. Still, if you want a good in-depth technical analysis of the physics of wood heat, the *Encyclopedia* is it.

Some of the more sophisticated stove makers have also run tests and their sales literature contains good information on their products' heat output. But none of this is going to help you much in choosing the stoves best for your needs. And here let's bring up a most important and usually ignored point: unless you hide it in your cellar, the wood stove will probably be the major article of furniture in your primary living space. You've got to like its looks; its size will have to match the scale of your room, its color your decor. People tend to fall in love with stoves for their aesthetic value more than anything else. Fortunately (within reasonable limits), you can fiddle around with stove installation and operation so that nearly any good stove can be made to heat you properly—with two provisos. Most important, it must be large enough. Better too big a stove than too small. But then don't fall for *too* big a stove. Every stove has a minimum draft requirement—will need a certain amount of air going through to keep the flue warm and fire going at all. In some of the big, super-efficient stoves that minimum can turn out enough heat to roast you out of a small room. Now, how to pick the stove or stoves that will turn out the heat you need where you want it?

The heating capacity of a stove is a direct function of the stove's size—both the amount of surface area it has to radiate heat and the pounds of material in it. You'll notice that older American cast-iron stoves are all numbered; the number refers to the weight, the pounds of cast iron in them. Till the oil crisis regenerated demand for stoves, and prices went skyrocketing, they were sold by weight too. A dollar a pound at the beginning of this decade, which was up from about a dime a pound back in the 1800s. These days, of course, you can't touch a good-quality, new cast-iron or steel and firebrick stove for less than a few hundred dollars, no matter what its weight.

How to figure the size and number of stoves to keep you warm? Sorry to get mildly technical, but heat is measured in British thermal units, or Btu's. One Btu doesn't amount to much: it's the amount of heat needed to raise one pound (pint) of water one degree Fahrenheit. Heating capacity of furnaces is calculated in the number of Btu's they can put out per hour. If you already have central heating and would like a rough idea of how many Btu's you are using, check the specification plate on your furnace. Gas heaters specify the Btu's put out per hour. Oil furnaces list an oil consumption rate. Multiply this by the number of Btu's in a gallon of #2 heating oil, 140,000, to get the overall heating capacity of your present system. If you've an all-electric heating system, I'm sure you are all too familiar with your Btu needs as you've tried to keep those astronomical bills down to something approaching a reasonable level.

## Figuring Heating Needs

The furnace that came with our own house has an oil-burning rate of 1.3 gallons an hour to heat both the house and the hot water supply. At that rate we would use, without the wood stoves, a bit less than 1,000 gallons a season. We figure the furnace would be burning at an average of a quarter of the time, year-round, but during extremely cold periods it would be going at full capacity. About a quarter of the energy goes into domestic hot water (based on the approximate relationship between hot water and space-heating costs in homes with families of our size that have separate water and heating systems). This would leave a gallon per heating hour or so to heat the space when temperatures were at their coldest, or 140,000 Btu's per hour. For our seven-room house that comes out to a convenient 20,000 Btu's maximum heat required per hour per room. From what little information we can find on space heaters, I gather that this 20,000 Btu per room figure is fairly accurate and typical. Not to say that you would be needing all that heat all the time, but to stay comfortable in all kinds of weather you should have about that much heating capability. And the figure applies to rooms up to twice the size of our parlor, 15-by-18 feet or 270 square feet on the floor with an eight-foot ceiling, or about 2,200 cubic feet. You might say you need five to ten Btu's per cubic foot of air to be heated depending on how warm you want it. These are *very* general figures, but as suggested earlier, trying to run wood heat figures out to several decimal points is an exercise in futility.

## Stove Size and Heating Capacity

In fiddling around with weights, surface area, general construction and heating capacity of a few dozen wood stoves, we've reached the following conclusions. For all types of iron, stone, and other massive stoves, heating capacity is a diminishing function of the stove's weight. A little stove, weighing in at 125 pounds, will easily heat one room of our well-insulated, storm-windowed parlor's size, but if you push it, or are very well-insulated, it will heat twice that amount of space. Double it in size, to 250 pounds, and it will only heat one more room; triple it, and you only get another room.

From our experience, a little iron hundred-pound stove, if of an efficient design, will hold heat and coals overnight, but can't hold enough to keep anything more than a small room luke-warm. Also, simply because of its smaller mass, it takes more wood than a big stove to generate a large amount of heat. On the other hand, because of the diminishing return on increased size, a three hundred-pound stove will not be three times as efficient a fuel user or heater as the little models. So, in our home we could have gotten about the same amount of heat from two one-hundred-plus-pound stoves as from the single three hundred-pound Combi. Plus another set of stovepipes to clean and ashes to clean up. The single big one is best for our square, relatively large-roomed house. Folks with other layouts might do best with a pair (or more) of one hundred-pounders, added chores in-cluded. Another advantage of the bigger stove, though, is its ability to hold a lot of wood. The bigger the stove, the more fuel it will hold, thus the more heat it can turn out between loadings. For example, on below-zero nights we can put in five or more six-inch diameter maple logs, close the damper down and adjust the draft so that each ten-pound log lasts about two hours. We'll get up to a good house-warming 30,000 Btu's an hour the night through, even if we all sleep in in the morning. A smaller stove, equally as efficient, would hold half the amount of wood and need to use it almost twice as fast to turn out that much heat. On the other hand, both the big and smaller models will produce half that amount of heat (about 15,000 Btu's) on half the amount of wood (some two-and-a-half pounds an hour), a fuel consump-tion rate that will keep the chill off a small house in not-too-cold weather all night long.

With the less efficient cast-iron potbellies and Franklin-style stoves where air input can not be strictly controlled, heat-ing capacity is also a diminishing function of weight. Our little fifty-pound kitchen stove back on the farm was fine for right there, but it didn't push heat into the drafty west end of the room. A larger stove, such as the Franklin in the homestead liv-ing room—in the two hundred-pound range—did a good job of heating several fair-sized rooms, as long as we stoked it round the clock, which we had to. But it had to be kept roaring away

*Franklin*

and on cold nights both wood consumed and sleep lost were great.

As something of an aside, let's not totally distain Franklin's and such other stoves lacking a full complement of baffles and gasket-lined airtight doors and all. As long as you have the wood to burn, proper dampers in the stovepipe and a well insulated house and/or a bed with a good supply of blankets, even a recreational Franklin will serve to heat a house in the coldest climates. Much of the state of Alaska is heated with oil drum stoves where airtightness and all that is unknown.

In welded steel and sheet metal stoves, you aren't so much interested in weight, but in surface area. Obviously the bigger the stove, the larger the fire and bed of coals it can hold and the more surface there is to radiate heat. One of the little sheet metal jobs you can get in any hardware store will put out your room's worth of heat, 20,000 Btu's, if you need it, but you have to keep a good fire going constantly. My calculations suggest that a steel stove should heat about half as many cubic feet of air as there are square inches of radiating surface on the stove. This is the same figure that applies to the largest cast-iron stoves, by the way. The smaller and more efficient sheet metal and welded steel models will heat perhaps half again as much, the number of cubic feet equivalent to 75 or 80 percent of the number of inches of radiating area. I guess what all this boils down to is that one hundred pounds of cast iron will heat two average-sized rooms if it's efficient, or only one if it's not. Each added two hundred pounds is worth two rooms in an efficient stove, but only one in a conventional one. And in steel stoves, the bigger they are, the more heat they'll throw. I don't know why I spent two days cussing over a slide rule and calculator to figure that out, but maybe it will help someone.

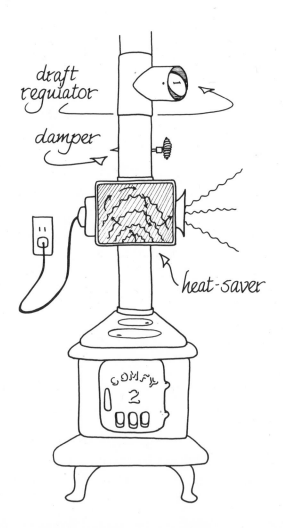

**Heat Savers**

While we're at it, now that we are into Btu's, we should mention the amount of heat you can rescue from the stovepipe with one of those heat-saver gadgets you must have seen advertised. They run flue gases around boxed, open tubes, then blow room air through the tubes. We'll discuss them in more detail later. The amount of heat they can pull depends on the temperature of air in the pipe. With a good hot fire, it can approach 1,000 degrees F., and at least one manufacturer claims its biggest unit can reclaim over 50,000 Btu's per hour at that temperature. I'll stick with the more conservative claims I've seen—from 10,000 to 20,000 Btu's per hour depending on temperature of the stack gases. Of course, the less efficient your stove, the more benefit you would get from the heat-saver. I hardly think they would be worth the money if yours is an efficient airtight-type design. It might even be detrimental, robbing the stack gases of heat needed to keep the flue drawing well. But with a big, roaring Franklin you may actually get more heat from the saver than from the stove. If we assume for convenience's sake that a saver saves an average of 14,000 Btu's per hour, in a 24-hour cold snap, it would retrieve 336,000 Btu's. If a gallon of oil is worth 140,000 Btu's, that comes to about two-and-a-half gallons of oil (selling just now for 50¢ per gallon) or a good dollar and a quarter a day. The unit should more than pay for itself in a single heating season.

## Buying a New-Design Stove

I've mentioned, more snidely than I should perhaps, the fact that stove stores are sprouting up everywhere and that all kinds of outlets that wouldn't stock so mundane an item as a wood stove five years ago are getting in on the action. You can buy stoves by mail order, through a dozen or more catalog sales operations, hardware stores, discount houses and big department stores as well as stove shops. In picking your own stove outlet you should ask yourself, and answer honestly whether you need help and information as well as a stove or two. Few of the large, more general retail stores and no catalog or mail order outlet can help you pick the right model or give either information or help in installation. But if you've done your homework and know what you want, you can often save money by purchasing from a high-volume retailer. Fortunately, in my view most manufacturers, importers and distributors of better quality stoves are pretty picky about the folks *they* will sell to, tending to pick knowledgeable wood heat specialists over the cut-rate discount houses. I suspect, though, that this selectivity will lessen in coming years as more firms enter the market and competition grows.

But for now, I find the best selections of the best quality stoves plus the best people with the most solid know-how in the small specialty stove shops. But not in all of them by any means. Just a few years ago most stove businesses were run on a shoe-string by what you'd call idealists—people to whom wood heat is an expression of preferred life style and a refutation of the oil-fueled values of much of modern society as much as a way to keep warm or make a living. But now that there is money to be made, and a lot of it, in stoves, the money men are moving in.

Not that I'm against anyone cashing in on a good thing. However, the stove seller who is in it mainly for a quick buck may not know anything more about wood than what he or she is told by the stove manufacturer's salesman. So, my advice is to check local newspapers for stove shop ads, then check out every store in your bailiwick. In general you'll find two kinds of operations and operators. The one to avoid—especially if you don't know precisely what you do want—is the place and people that will sell you anything you want. You see stove "boutiques" opening up now, complete with piped-in music, carpeted floors and indirectly lighted product displays. They turn me off too. I'd advise the novice woodburner to seek out a genuine stove-person—someone with stove polish in his or her blood, who knows the trade from personal experience and has been in the business for a few years. (Wood heat being the recent phenomenon it is, a stove shop that is three or so years old is an old-timer.)

Your better salesperson (like ethical salespeople in any business, I feel) will show an interest in your particular heating needs, will try to help work out a combination of stove and installation that is best for you, and not try to move you toward the biggest ticket item on the floor. I would personally favor an outfit that also installs stoves and flues (though you'd best be sure they know their stuff, as mentioned in the flue chapter). Another criterion is whether the shop people heat with wood themselves, both at home and in the store. This is too much to ask of every person in every outlet in the middle of a city—and wood heat has its place in town as well as the country. But you will get the best advice, best selection of products from people that have had a genuine, personal commitment to wood-heated living for some years.

### Kinds of Stoves

So far I've concentrated primarily on the new stove designs we see most often in advertisements—those with baffles and airtight doors and newfangled draft systems. But let's reexamine the Franklins, potbellies, log burners and other cast-iron stoves that I guess we can call conventional designs these days. As mentioned earlier, the Franklin fireplace is a far remove from Ben Franklin's original design—it has evolved into a type of freestanding fireplace, and when installed with a prefab metal flue, is a way to have a cheery grate fire without the cost of putting in an expensive brick fireplace. Most Franklins come with a log or coal basket, so that air can come in under the coal bed as well as over the top of the wood, providing an ample oxygen supply, thus a good, if comparatively wasteful blaze. A damper placed in the smoke hole of the stove or, in more primitive models such as the one we had on the homestead, in the stove-pipe leading from stove to flue, gives you considerable control of draft. Most also have closing doors, usually with small closeable draft openings in them. Having heated for a good decade with an antique Franklin, and in a frost pocket of snowy New England at that, we can attest that they will keep you warm. There is a smoke shelf built into a good Franklin, as we'll discuss in detail when we get to fireplaces, and it serves much as the smoke baffle in the Scandinavian designs. You do have to stoke a Franklin every four hours or so to keep warm in a drafty house and they use more wood than the more efficient designs. But they heat.

The same is true of potbellies and other conventional stoves. Indeed, a properly run potbelly is as effective a heater as you could ask for. It will have a draft opening under the grate to feed oxygen to the coals plus another in the belly to provide air for secondary flame combustion, so a good one can really put out

"pot belly"

the heat. You'll find this same draft configuration in most other conventional iron stoves: the log burners, Fatso-type laundry stoves, et cetera. The main difference between them and the more popular airtights is the lack of precision in fitting parts, doors and all, so they are leaky, they burn wood faster, and they won't hold a fire as long.

In a sense, these stoves, or most of them, are examples of the last gasp of a stove industry that was nearly killed by petroleum. By the 1920s, there simply wasn't the demand for ornate, exactingly made (and airtight, baffled, efficient, beautiful and expensive) parlor heaters or cook stoves that there had been in the pre-oil era. About all that would sell were what I'd call occasional heaters used to warm an isolated cabin during deer hunting season, the ornamental Franklins and the one truly efficient design that seemed to appeal to modern sensibilities, the cabinet-style circulators.

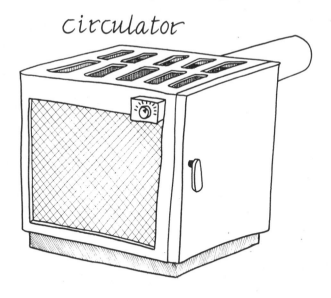

circulator

## Circulating Stoves

Circulating wood heaters come in contemporary-styled cabinets and have automatic thermostats, patented downdrafts and are as capable as any stoves made. One advantage is the (usually) optional fan that blows air out and circulates it into the house. With the blower, most manufacturers claim their units will heat an average-sized house, and I know plenty of New England homes that have been heated with circulators for years. However, I admit to an utterly unjustified bias and wouldn't have one in the house. Reason? The fuel doors are on top or at one side and you can't see the fire, which you can by opening the front-located door of most radiating stoves. You can't cook on them since the cabinet never gets warm enough to even heat up winter-chilled feet. You never even smell smoke, they are so carefully engineered. It's the same as having an oil furnace in the room. The only difference is you have to load it with logs once in a while, and empty the ash tray. I must admit, though, that a lot of folks would find these features positive advantages.

Just as with conventional central heating, room size, type and amount of insulation and siding, whether or not you have storm sashes and doors all enter into the equation of how many to get, what size and so on. So, get the big model circulator for a big house, a little one for a little house, set the temperature you want and keep adding wood. It's the only way to have wood heat without a lot of experimenting and not a little shivering and sweating. But no fun either, the way I see it.

Still, a great advantage of the circulators is the near-automatic operation. They are probably the safest designs of all too; a spark simply can't get out of one, and if you keep the automatic damper mechanism in good condition they can never overheat. As stated above, the outsides do not get hot enough to burn, which is the case with any radiating stove. If there are babies in the house, particularly if you've more than one crawler or toddler and can't keep an eye on all at once, a circulator can save

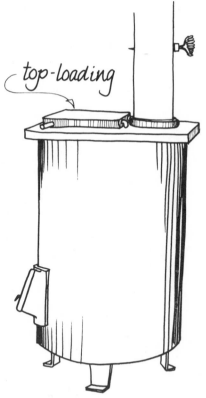

top-loading

cylindrical, steel plate
("Shenandoah")

you a lot of worry and potential grief. The metal of a well-fired radiating stove can take the skin right off a little hand, and don't forget it. Kids do get wary of the stoves in time, and at a surprisingly young age. But until they do, a newly installed stove is a hazard—except for the cabinet-enclosed circulators.

Most circulators have firebrick linings, cast-iron grates and sheet metal outer cabinets and will last forever. Though each brand has its own method of introducing air for combustion to the fire and blowing room air around the hot interior and into the room, some of which are patented, the principles are pretty similar. (One of the best and widely known brands, Ashley, has a patented downdraft system for warming combustion air. I don't suggest that they are attempting to mislead anyone for a second, but this is not the kind of downdraft Ben Franklin tried in his coal stove—and that is most successfully employed in central heating furnaces to force gases produced by the fire back into the coals or flame so fuel will burn as completely and economically as possible.) Here is a generalized description of the way circulators operate.

The firebox usually has fire door and ash removal from one side, smoke hole in back and thermostat in front. Air for combustion normally enters at two levels. Air for primary combus-

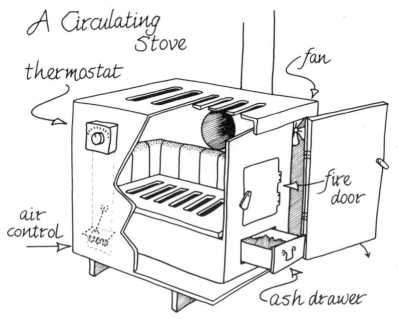

A Circulating Stove

thermostat

fan

air control

fire door

ash drawer

Automatic damper mechanism:

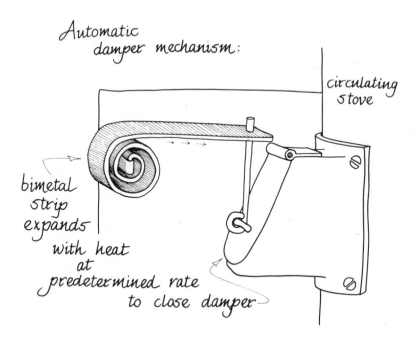

circulating stove

bimetal strip expands with heat at predetermined rate to close damper

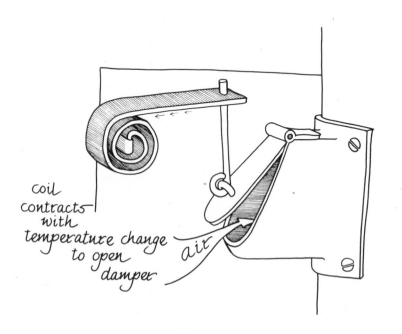

coil contracts with temperature change to open damper

air

tion at the level of coals enters at the bottom, while air for secondary combustion of gases enters at the top. The amount of air entering is regulated automatically by a bimetal strip—two different metals fused together in a coil or other spring shape. As they are heated (almost entirely by warmth radiating out of the stove—not by room air as popularly thought), they expand at different rates and the stress causes the strip to expand at a predetermined and constant rate. With experience you can set the thermostat to keep room air at a constant heat. The strip will bend back and forth, closing or opening the draft control to maintain a fire that will keep the heated air flowing past it just at the temperature you've set.

Room air is forced around the hot inner firebox by gravity or a fan attachment, all very automatic. Some brands have complicated smoke aprons and ash screens that close when the cabinet and firebox are opened so you never smell smoke or see a flame or get ashes on the light beige wall-to-wall carpet. As I say, I wouldn't have one in the house. Louise wouldn't have light beige wall-to-wall carpet either. But if you do, only a circulator will let you keep it clean without a lot of extra precautions. A final note: Be prepared to replace the thermostat mechanism

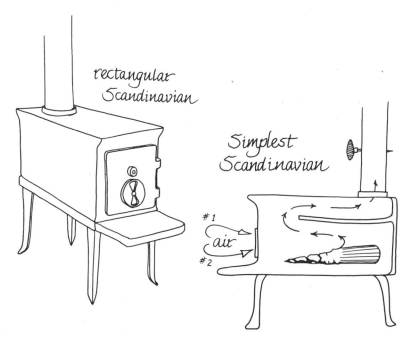

every few years. The bimetal strip and attachments simply wear out, a fact that some stove sellers either don't know or don't tell you.

### Scandinavian-Design Airtights

Now for the most popular designs on the current market, the cast-iron airtight stoves made in Europe for many years plus a whole rash of out and out copies such as the Reginald box stove imported from Ireland and a growing number of North American, Korean and Taiwanese imitations that have importers of the originals mad as a hornet whose nest is under a cow. The widely publicized "new secret" of these stoves, now incorporated into nearly every design to come along, but an idea originating hundreds of years ago is complete airtightness and one or another system of interior baffles that force the smoke into a serpentine pattern of movement inside the stove. The simplest is a single plate extending from the back of the firebox to about three-quarters of the way to the front. It's advertised

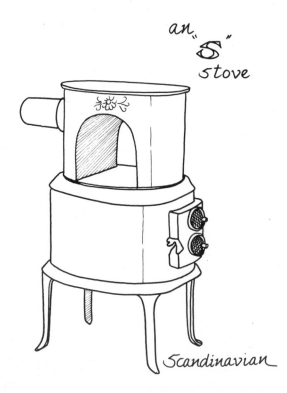

an "S" stove

Scandinavian

that in these stoves wood burns "like a cigarette" from front of the stove to the rear, and it does. The only air that is admitted comes in at two levels from the finely machined draft control for primary (1) and secondary (2) combustion.

All Scandinavian airtights and some of the copies are beautifully cast and assembled. The door and opening faces are either finely ground or fitted with asbestos rope gaskets and all have latches that close them securely. Firebrick or iron side plates around the firebed make for a longer life. They come in standard black iron that needs occasional "polishing" with stove black, in black fireproof enamel and in an assortment of other colors. Jotul offers a deep forest green and other brands come in unfamiliar, perhaps, but gorgeous shades of rich red, blue and other colors. The enamel is pretty tough but can chip. We prefer flat black in an iron stove because a coat of stove polish can repair any discoloration. The plain ones are cheaper, too.

If there is a disadvantage to the genuine imports it is cost—they *are* expensive. And, since most are manufactured in coun-

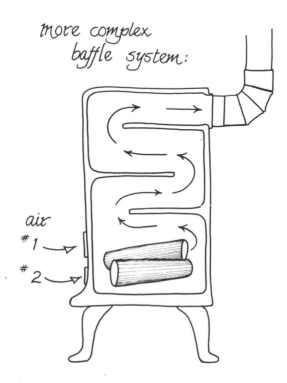

more complex baffle system:

air
#1
#2

two inches
of
sand
in
bottom

tries using the metric system any replacement nuts for provided stove bolts must be metric sizes. So must the stovepipe, or at least the section fitting to the metric-sized smoke boot. You can purchase metric-to-inch adapters, or improvise. In adapting metric-to-inch pipe or any other change in pipe size always go up in size—get the nearest larger size, never smaller.

Unlike most native U.S., Canadian and Central European designs, the typical Scandinavian stove lacks a grate or log rests. The manufacturers recommend building up a deep ash bed to protect the bottom from burning out, but I'd suggest putting at least a two-inch layer of sand in the bottom of any grateless stove at first. Particularly if it's an expensive import.

## North American Airtights

Now the imports are lovely things, and for a few years were the only efficient radiating-type airtight heating units we could get. But as popularity of wood heat has grown, so has the number of North American-made stoves. As said elsewhere, anyone with a welding outfit can slap a stove together, file down weak and sloppy welds, paint it black and make outrageous claims about the product's uniqueness. There are no industry-wide construction or performance standards for stoves, no really objective tests or comparative data available, so you are on your own in selecting your own wood heater. However, taking a cue from the European stove makers, more and more reputable domestic manufacturers are turning out brick-lined, airtight, efficient stoves.

Particularly if you are buying a stove with an automatic damper system, a blower motor or any moving parts that may need service or replacement in time, I'd think twice before spending a lot of dough on a stove from a new, probably under capitalized and poorly managed firm that may go broke during the next relatively warm winter. Such long-established concerns as the Washington Stove Works or the Atlanta Stove Works (listed in the Sources at the end of the book) have been around for a century or more and will likely be here to supply parts for at least a century to come.

The main disadvantage I see, however, in sticking to the big, long-established firms is that they are not leaders in innovative design. (If you can sell all the eighteenth century-design stoves you can make, why gamble on a chancy new idea?) So we're seeing the really novel ideas coming from new entries in the field. And not all new designs are poor, not all new manufacturers incompetent or bad managers.

One of the more interesting designs to come along is the cast iron Defiant and its smaller version, the Vigilant, made by Ver-

Defiant

mont Castings, Inc. (which is not a foundry as is more or less implied, but is an assembly and sales operation). The designer managed to combine every popular design feature into a single stove. You can load from front or side; there is a complex baffle system, firebrick liner, complete airtightness and an automatic thermostat; you can cook on the top; you can remove the front doors and put on a screen to see the fire. It has a nice colonial appearance, resembling the good old Franklin. The stove does put out the heat and is almost as well finished as one of the 1890s masterpieces. Only problem is that to clean it out, you have to break the stove—you have to knock out a cast iron door at one side that's sealed up with stove cement. The directions say it swings aside, but owners I know are afraid to try to get it open for fear of breaking the brittle cast-iron casting. The designer is an ex-architect and knows his design, that's for sure, and some people call his stoves the Cadillac of the industry. But I'd be darned sure I could clean it or any other stove before shelling out over $500 for it.

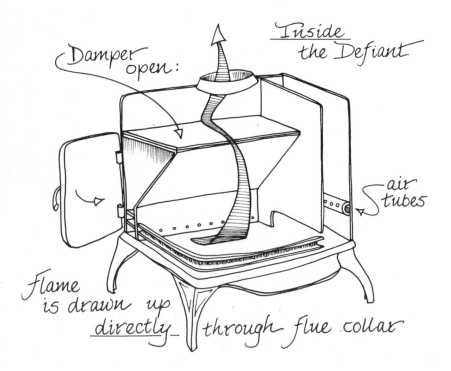

Damper open:   Inside the Defiant

air tubes

flame is drawn up directly through flue collar

Popularity of the Defiant has generated copies, just as with the Scandinavian designs. One model designated the Upland #207 repeats many of the outward design features of the original, including airtightness, but has eliminated most of the complex interior baffling and much of the cost. Only time and experience will tell, but for the typical American woodburner, I'd bet that the simpler design will prove the winner.

There will be other copies too from the busy foundries in the Far East. Korea, Taiwan and who knows what other nation's iron foundries are already turning out sleazy copies of any number of iron stoves, new and old designs. The best ones I've seen are mediocre reproductions of some interesting antique 1860s-era transitional stoves—in a period when they were adding flat cooking surfaces out in front of what is essentially a heater. Worst are little mini-potbellies with doors having only one hinge. So far, all the Oriental iron we've run across isn't worth the gas to drive it home: thin and pitted castings, poorly mated and finished. Junk.

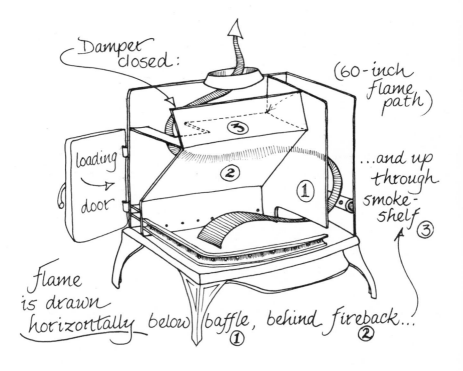

Damper closed:

(60-inch flame path)

loading door

...and up through smoke-shelf ③

flame is drawn horizontally below baffle, behind fireback...

## Welded Steel Stoves

The easiest way to make up a stove is to cut up hunks of boiler plate and weld them together. Currently there are hundreds of shops doing just that. But here again, I fear we are beginning to experience the "talented amateur" problem; just because a fellow is a good welder doesn't mean he can design a good wood stove. And, since welding—or high-temperature welding of steel—is a comparatively recent development, we lack any old, time-proven examples of welded stoves. You'll see quite a few older stoves framed in strap iron, with body plates of iron, soapstone or whatever. As we'll mention further on, they were held together in many ways, usually by stove bolts and furnace cement. But none were welded. Couldn't be done back in the heyday of the wood stove industry.

Now, there's no technical reason I know of why a stove made of sufficiently thick steel can't hold a fire if the welds are sound. Problem is, many amateur welders are turning them out and I know one stoveman who has to turn away one out of four stoves of a well-accepted name brand just because of defective welds. A worse problem in my view lies in stove design itself. As said elsewhere, innovation for innovation's sake seems to be the guiding principle in today's stove design. And not all stove designs are proving out.

For example, the most widely copied new design is the Fisher, thought up by a welder in Oregon a few years ago and now being made in several locations around the continent. As the illustration shows, the major design feature is the stepped top—a sort of shallow S shape that is hotter at front than in back, thus providing two cooking heats. The Fisher, now in several sizes and designs, has a firebrick-lined firebed, massive cast-iron airtight door and none of the automatic thermostats, baffles and all that complicate operation and maintenance in more complex designs. Well, the Fisher was one of the first domestic designs to hit the market and the stepped top was copied by dozens of shops, large and small. However, we are now learning that some step stoves are beginning to warp, ruining their lines, stability and threatening the integrity of their welds. The problem, I speculate, is that double-curved top—the stresses and im- balances in temper introduced in the bending process produce

a
Fisher,
fire-brick
lined
stove

differential expansion as the stove is continually heated and cooled. In time, these differences may produce a warp that will go on till the stove is all skewed out of shape. It will be several more years, though, before we know for sure (if we ever do), which step stoves are good and which are the bummers. I hope the design itself doesn't prove to be altogether bad. Fisher and most of its copies are nicely made, good heaters and have set a lot of folks back several hundred dollars each.

One innovation originally put on the step stove but now found in several other outer configurations is a series of steel tubes which run around, through, under or over the firebox. There's a blower attached at the side or bottom where the tube network begins, and air is blown out where the tubing is welded to the top side or front of the stove. Here once more, a good sounding idea not proven over time, and at least two serious potential problems have developed. One is that the in-firebox tubing in some models was so thin that it is beginning to burn through. One major manufacturer of this design has increased the tube thickness by 100 percent, and as a disillusioned stove shop operator told me, "They didn't do that for the fun of it— the original is a menace . . . all you need is that first pinhole from a combination of rust and burn out and poof! Up goes your house." Fortunately, this fellow only sold a half dozen of the stoves before he found out the problem, and he had the integrity to call them in for refund or replacement.

The other problem is still speculative, but several met-allurgists have questioned the integrity of the welds in these stoves. In effect, the makers have tried to fuse three or more dif-ferent metal alloys—mild steel boiler plate, a cast or rolled steel tube network, plus the welding rod itself—into a whole. Then they are exposing the lot of it to different stresses under firing: two temperatures on the top tending to twist or expand at dif-ferent rates, the tubes with air going through them, making the steel hot on the outside, cooler inside, plus who knows what dif-ferences throughout the system, all tending to break the stove. One old welder who worked on railroads in the days of steam engines—when they made boilers that had to last—predicts a life of no more than three years for these stoves in sustained use. Well, they haven't been out for three years yet and some brands claim a "lifetime" guarantee in their ads. Believe whom you wish. But I for one would wait a while before buying a welded steel stove, particularly if it has an air-pipe system or some sort of compound curve in the design. Haven't heard anything yet against the simpler box-design steel stoves, but I'm not per-sonally convinced that welded steel is the best stove material. I'll stick with cast iron, at least for the coal bed, in any stove meant for continual use.

Not everyone agrees, of course, and the final word is by no means in. Several knowledgeable stovemen I've talked with stick by the steel stoves. But they favor the simpler designs, plain cubical or shoe-box style stoves. One fellow uses a big Sierra—a variation on the step stove, but with the top made of three separate plates welded together, sides reinforced with angle iron—to heat his shop and home. He put in a baffle plate on his own hook, and though the stove lacks a gasketed door, it is airtight enough to hold a fire overnight and then some.

An intriguing new design, developed by Eric Darnell of Sharon, Vermont, consists of parenthesis-shaped lengths of tub-ing welded together right-hand curve, left-hand curve, right-hand curve, and so on to form a cylindrical firebox. Airtight ends with a smoke hole in back and a front door are welded on. When the stove is fired up, cool air from the floor is drawn up through the tubes, heated as it rises. The tubing exposes a curved inner surface to the fire and directs hot airflow toward the ceiling, in theory at least putting out more heat and generating more air

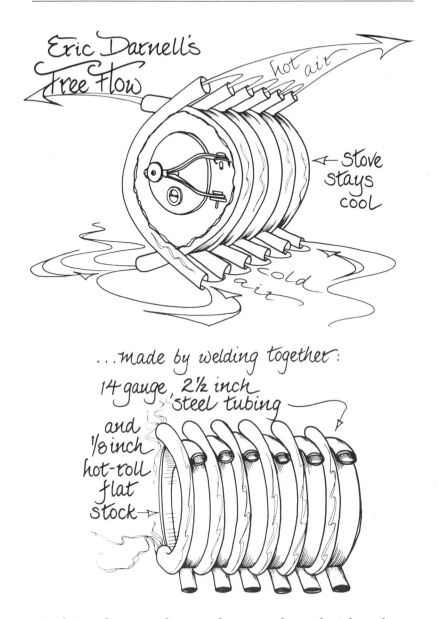

Eric Darnell's
Free Flow

hot air

← stove stays cool

cold air

...made by welding together:
14 gauge, 2½ inch steel tubing
and ⅛ inch hot-roll flat stock →

circulation than an ordinary radiating surface. This idea is beautiful in its simplicity. But again, how will Darnell's welds look in ten years? Have to wait.

And the innovations keep coming on. The S.E.V.C.A. stove illustrated is one ready-made version of the dual-drum stove we have illustrated in the section on building your own heater. It

has a baffle system and above the main firebox, a half-drum-shaped secondary heat exchanger that extracts so much added heat from the smoke that it rated highest among new stove designs in one test. Problem is, it's made out of old propane tanks and, though I'm all in favor of recycling, I wonder how long an old propane tank will hold together with a hot fire in its belly. Also, cleaning it—the top part and baffle system in particular—must require a contortionist.

Still another idea being introduced in steel stove designs—or being reintroduced, as it is hundreds of years old—is the downdraft principle. You may recall that Ben Franklin tried this out a while back, and it failed. Today's designs are less ambitious, don't try to recycle the smoke, but pull air in at the base of the fire, so oxygen is blown right over the coal bed or into the flame so that theoretically you get greater combustion and less smoke and creosote buildup.

One downdraft design is typified by the Tempwood and imitators. A simple welded steel cube, it has a round hole with a cover and two tubes with moveable lids that aim down into the firebed fixed on top. Air is heated as it is drawn in and down to the fire and a common claim is that you can "see the blue flame of secondary combustion" at the base of the air entries. Well, OK. The stoves lack doors, are simply made, so relatively good bargains, heat well and should hold up over time if any welded steel design will. They are also uniquely attractive in an austere way. The only hassle is in loading and cleaning. Your only entré is the foot or so wide hole in top. No really big logs will fit in and cleaning takes a little scoop with a long handle. I don't see how you could do any cleaning unless you let the fire die and the house get cold, lest you singe your hide.

Another downdraft principle is typified by the Vermont Downdrafter, innards illustrated. Here the secret is a V-shaped grate with a fairly sophisticated air and fuel delivery system and a built-in heat exchanger in the top of the firebox. The stove has loading and cleanout doors, and a simple, hopefully stress-free box design. It also contains more of the sophisticated draft features of the 1890s stoves than any other current domestic design. This surely biases me, as I've admitted a preference for the time-tested ideas our forefathers lived with for generations.

Still and all, if asked to recommend a modern wood stove, I vote for cast iron held together with bolts and cement. You can disagree, and you are welcome to; you're in good company. But, they build highways in blocks, leaving tar-filled joints every so often to absorb expansion and contraction, and I feel (just now) that a heavy-duty stove should be made the same way. Weld it

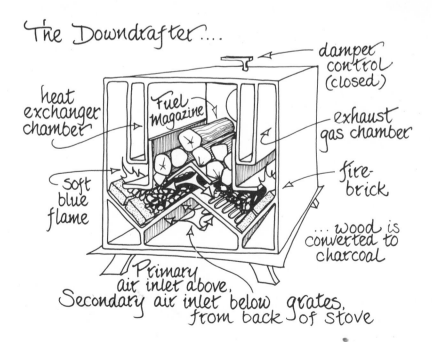

The Downdrafter....

damper control (closed)

heat exchanger chamber

Fuel magazine

exhaust gas chamber

fire-brick

soft blue flame

... wood is converted to charcoal

Primary air inlet above.
Secondary air inlet below grates,
from back of stove

up and the stresses may come out in warps—*may* now; the final tally isn't in, as admitted before. But if I were buying a welded airtight steel stove today, it would be a four-square cheap design that I could discard without regret if a few years' use proves it a bad design.

### Creosote

Before we leave the airtights, imports, domestic or home-made of whatever design, we'd best mention their greatest drawback, and this applies to radiating stoves, circulators—all of them, in spades! When closed down, the fire is flameless. However, because of the very limited amount of oxygen entering the firebox, combustion may be incomplete. Several components of the wood vaporize without burning in the intense heat and escape into the flue. Upon reaching the relatively cold stovepipe or flue liner, they condense into a fluid called pyroligneous acid. This is a brown sticky substance consisting primarily of methanol (wood alcohol) and acetic acid, the same stuff that gives vinegar its bite. It can drip from your stovepipe and stain the carpet, but is otherwise harmless in itself. However, the acid tends to trap carbon in smoke, dry and bake on the insides of pipes and flues, becoming what old-timers call creosote. If you're interested, the acid is one component of the commercial creosote you buy for such things as preserving wood. Commercial creosote is obtained by distilling wood, beech in particular, by heating it in the absence of flame to the point that the several aromatic oils in the wood turn to vapor; they are pulled off and cooled, whereupon they condense into creosote. This is precisely what can happen to part of the wood in the flameless firebox of your airtight stove; you are distilling the logs at the top, the ones up and away from the burning coals.

### *The Danger*

Once dried out, creosote is flammable stuff, though it takes a lot of heat to set it afire. I've seen plenty of antique flues with many inches of creosote buildup on them. The relatively inefficient fires of old damperless fireplaces had a comparatively hard time getting up to a creosote-tindering temperature. But plenty of them have done so; the dozens of cellar holes scattered through our New England woods attest to that. However, the danger of a flue fire is minimized if you have a modern chimney with sound brickwork and liner. Chances are the flue fire could burn itself out without destroying the house, though the great fountain of flames showering the neighborhood with sparks will bring the fire department in a hurry. In an older flue with cracks that might let fire out and into the house timbers, a flue fire can be a disaster. In any event, you should avoid them if possible and have that fire extinguisher mentioned earlier good and handy to control one, if and when.

### Beating Creosote Buildup

All airtight stoves will produce creosote if operated at maximum efficiency in the closed-down mode. Frankly, it scares me to see so many friends putting them into the old flues of antique homes; if I have any say in it, they get an experienced mason to go over the chimney from ground to roof before they light up. And no one should attach an airtight to an extended length of stovepipe. One of our photographs shows the loose creosote accumulated in the pipe of an old turn-of-the-century airtight in our house. Had that creosote ever caught, the tremendous demand for draft would have sent the pipe into gyrations; it would have torn itself apart, hurling chunks of flaming creosote all over. Long pipes are okay for some installations, but not for today's airtight stoves. Plan to vent yours directly into a permanent masonry or prefab flue; minimize stovepipe.

Now considerable creosote buildup can be avoided by having an open fire during most of each day. This is one reason we chose the Combi for our main wood heater. It is kept open during the day, a low fire going most of the time, with the damper nearly closed. Then we pile on the wood for a high-flame cheery fire during the evening. The stove is closed up for most efficient, (but creosote-producing) heat only during the night or when we are away.

There are several products on the market that purport to remove creosote from your flue. They go by such names as Fluesweeper. First of all, they are nothing but rock salt that you can buy by the one hundred-pound sack for loose change. And they work by vaporizing the salt, which combines with the creosote to make it catch fire. Perhaps if you threw a handful of salt on the fire each day and burned the creosote as it accumulated, it would be safe. However, no government agency or fire prevention association that we contacted recommends using the stuff, so neither will we. If you are going to install an airtight stove, do your planning as if a flue fire is inevitable, then do your best to avoid having one by cleaning the flue at least once a year. And, during the day, have an open fire when you can—draft fully open.

### Flue Fire Alarm

While we're on the topic, you'll find no commercial flue fire alarm on the market. But a friend happened upon this dandy alarm in an old book of farming information: "Drive a nail in two rafters on a line with the face of the chimney, to which stretch a cord close to the chimney, so that, in case of fire, the cord will burn off and release the weight hanging to it, which in turn will

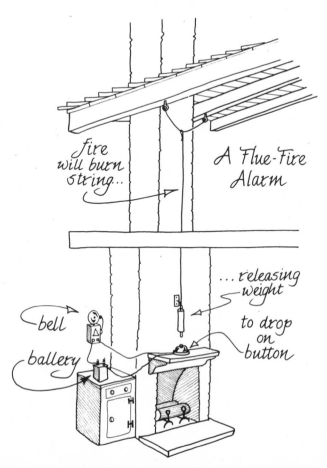

fire will burn string...

A Flue-Fire Alarm

...releasing weight

to drop on button

bell

battery

drop on an electric button and ring a bell. A dry battery will cost 20 cents and a bell 50 cents. Place these on a shelf above the fireplace. Place a piece of heavy wire, ten inches long, as shown, and fasten to the wall or chimney for the weight to slide on. The weight need be suspended only an inch or two above the bell."

## Buying an Old Stove

Probably the best advice to give on the topic headlined above is "Don't." For one thing you will probably have to pay a premium for the age factor, and just about any old stove will be used—really used. In the old days they fired stoves till they literally fell apart. Most spent at least the latter part of their first go-round burning coal, which generates a heat intense enough to burn out even cast iron in time. Further, no single book on earth can list the names of the huge variety of stoves that have

ornate, parlor

been manufactured over the years, to say nothing of telling you how to run them or what parts they need (and which may be missing). However, everyone would rather have one of the old-timers if he can find a good one (we have six at present count) so, here's what I'd look for in the used stove shops.

**Inspecting the Goods**

First, don't be afraid to probe and pry, open doors and poke around on an old stove, even if you come out with your clothes covered with stove black. You should get to know the insides, the damper systems and smoke channels if any, as well as you do the outside. As I've said, the old-time stove designers came up with an endless number of innovations and a good many stoves are one-of-a-kind. Chances are the antique dealer or stove shop operator won't know any more about the stove than you do. (Here and there you will find a stove person who has taken the time to bone up on old stove design and restoration. Good ones are rare, though, and you'd best be prepared to be your own expert.)

The size to heating capacity relationships cited earlier apply to old as well as new stoves. You'll find that really large stoves, capable of heating the big open sort of homes we tend to favor these days are hard to come by. Back in the heyday of stove design, people used stoves just as they had the colonial fire-places—a smallish one in each living area and a big one for heating and cooking in the kitchen. If you do find a really large antique heating stove, chances are it was used to heat a factory, store, school room or other large area. The big oldies are often fairly plain and utilitarian in looks, but they do the job.

Most of the artistry in old heating stoves, the fancy bas relief scrollwork and nickel plate is found in the parlor heaters, usually two to three feet high and two- to three-hundred pounds in weight. They were usually made almost completely of cast iron, many with finely ground, airtight doors and multiple draft controls. The several castings were hand-fitted, cemented and bolted together, just as in the better modern iron stoves. Most will date from the late 1800s to the 1920s and will have grates to permit air to enter under the fire; earlier stoves, now properly classed (and priced) as antiques, lacked the grate. Indeed, those much copied modern Scandinavian designs are similar to stoves made in the late 1700s and are actually primitive designs by the standards of the 1890s.

Probably the ultimate in stove designs are the columnar or cylindrical stoves of the late 1800s. These have a cast-iron firebox, top, doors and smoke outlet. But the high, tubular body is usually made from two sheets of metal, pink "Russian iron" being the most popular. The sheet metal was crimped and/or riveted together at the back of the stove; the inner liner was designed so it could be replaced in time when it burned through. Some of the smallish tubular stoves you find will be firebrick lined at the base. These were coal stoves, used by the thousands to heat railroad cars, trolleys and such, though they will serve just fine to heat your living room with wood.

If you are lucky, you'll run into an old illuminated stove. These had isinglass windows in the doors, and up to a dozen doors all around so you could see the fire. Devils to keep clean, though, as the inside of the isinglass smokes up quickly. It also is easy to break, cracks and falls apart in time and is not all that easy to replace.

Many of the old sheet metal stoves and a few all iron models have complex smoke channels, heat exchangers, built in or attached in a separate unit at the back. These ingenious devices run the smoke around much as if they had bent a smoke pipe double or in a series of flat S curves, all to squeeze heat out of the stack gases. A few of the more sophisticated designs include a smoke recycling system; a portion of the air for combustion is introduced in such a way that it pulls some of the smoke down through a channel in the back of the stove and in under the coals so it can run back up and burn.

I've seen stoves with parts numbering in the hundreds, a dozen or so draft, grate and smoke controls, isinglass doors, bowls in the top to hold water to humidify the air, cooking lids in the smoke boot, rollers on the legs so you can shut them up and haul heat around with you and just about everything but a built-in TV set. (Of course, there was no TV in wood heat's first go-round, but toward its end radio had come along. And there is a 1920s-era stove designed like those big old domed wood veneer radios, complete with a fake grille and nobs.) So, be prepared for some interesting surprises in the old-stove shops. If you are lucky, you may find a stove that is airtight, baffled and as efficient a heater as any modern design—maybe more so. Remember, there weren't any chain saws in 1880 and those old stoves were designed to get the most heat possible out of the wood. Indeed, some I've seen are perhaps too efficient, rigged with so many control and draft and smoke manipulators that it is almost a full-time job to run one.

Also, cleaning the more sophisticated old stoves is complex, requiring a good supply of long-handled brushes and ash rakes. I've never seen an old stove that *had* to be dismantled for cleaning, but many should. In buying your old stove, bear in mind that our forefathers dismantled their stoves every spring, brushed and blacked them, reassembling them for airtightness with new stove cement and greased bolts. So, unless you've the skill and inclination to do the same (and few of us moderns do, myself included I'll admit), steer clear of the fancier old designs and buy one you can live with on your own terms.

Our house in town came equipped with a couple of old wood-burning stoves, one of which was this 1898 Cole's Airtight. By today's standards, installation wasn't safe, as the stovepipe entered the flue too close to the ceiling and was too loosely wired. But it heated a part of the house for more than 75 years, run by a fellow who learned wood heating a century ago and knew how to keep the fire in bounds. We decided to rejuvenate the stove and install it in our basement workshop.

The first step was to take down the stovepipe, which had several years' accumulation of soot in it, and move the stove to the outdoors for disassembly. Many rusted soft-iron stovebolts had to be cut with a chisel, hammering gently to avoid splitting the old iron. The soft iron liner was burned through and came right out. The brightwork wasn't at all bright, so it was sent to a chrome/nickel electroplating shop—the sort that trades with antique auto buffs—for replating.

After the stove was lined in galvanized steel at a sheetmetal shop, I lined the firebox with soft but heat-impervious K-26 firebrick. The bricks I carefully trimmed to fit, then cemented in place with refractory cement. The top and the brightwork went back on, the bottom was covered with several inches of sand, and the stove was ready to hook to a flue. It all sounds easy, but it took a lot of cutting, drilling, punching, fitting and cussing to get everything together airtight.

With the stove at the ready, I next chiseled a hole in the flue in the basement and cemented a thimble in place for our "new" basement heater. The best approach here is to chip out most of the mortar surrounding one center-course brick. Then split the brick and remove it. Finally, chisel away at the adjoining bricks until you have the right-sized hole for your thimble. The thimble is an adjustable metal sleeve that goes in the flue, making installation and removal of the stovepipe easy. And since the Cole's Airtight is positioned in front of the chimney cleanout, ease of stovepipe removal is significant. It was a whole lot of work, but with the old girl newly blacked, polished, installed and fired up, it almost seems worth it. I'd bet she'll last another 75 years.

## Bonding Breaks, Patching Holes

I can't think of much about buying old sheet metal stoves that isn't covered in the repair job on the Coles airtight shown in the photos. Look for rust inside and out. Be ready to replace the liner if there is one, and to patch holes if there are any. You will have to use nickel welding rods and patching metal capable of withstanding at least 1,000 degree F. temperatures. A welding shop has the equipment and know-how to handle it, and since I don't, I'd take any patching jobs to the pros. If you are one yourself, fine. But patches put on with the little propane torches (or even the new hotter-firing welding gases such as BernzOmatic's Mapp Gas, which I use around the house) could come unstuck on a hot stove. Nickel welds will hold, but I would never try patching a sheet metal stove myself. I wouldn't trust the mend to hold up under all eventualities, such as overheating, and I wouldn't feel confident with the stove. Put patches on with equipment capable of cutting through boiler plate. And, frankly, sheet metal stoves were never very costly or well-made things, even in the old days. Unless you find a genuine rarity, I'd say that extensive patching of an old sheet metal stove isn't worth the time and effort. Besides, if it is old enough to have an antiquarian value, the metal just may be too corroded or soft or flimsy to stand up to the heat needed to apply a safe patch.

Cast-iron stoves are something else. Some of the old beauties are worth a lot of work. But one warning: you can not rebond broken cast iron so it will hold a fire. I don't care what anyone tells you. Sure, there are modern space-age glues and welding techniques that will mend a break so a stove leg, say, will hold weight and look fixed, but it isn't. You see, as it heats up, cast iron expands at a rate that it feels comfortable with. Nothing else expands at quite the same rate, and if you try to put a layer of glue or welding bond between one piece of cast iron and another, the three pieces will expand at three different rates and you have a break all over again. The only way to repair cast iron to really last is to replace the entire casting. Some stoves are common enough that you may be able to find two or three with good salvageable parts to go together in one good one. Or for a price you can have a foundry cast you a new part. You'll probably have to look and talk hard. Few foundries are in the

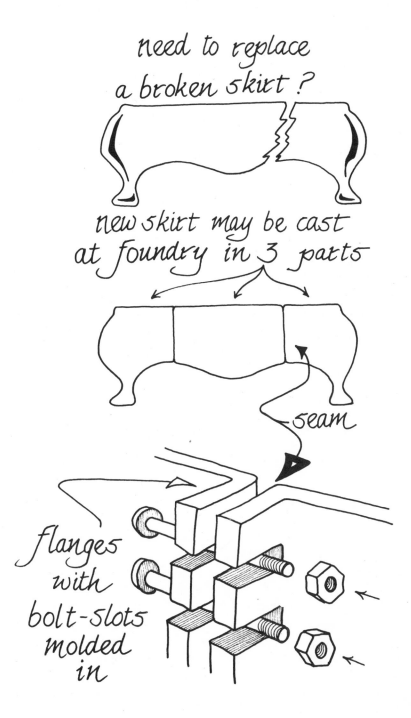

need to replace
a broken skirt?

new skirt may be cast
at foundry in 3 parts

seam

flanges
with
bolt-slots
molded
in

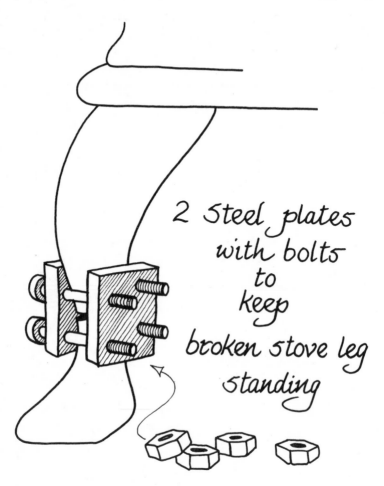

2 Steel plates
with bolts
to
keep
broken stove leg
standing

small-job business. Many no longer use the split-box, sand casting technique needed for most custom work. But do look and talk. And be prepared to shell out. Take the broken part to the foundry boss for instructions on how to make a pattern for their mold. You may find that they will prefer to break up a large casting into several parts. You'll have to work up the patterns in wood, clay or plaster of paris. It's a lot of time and effort, but worth it. Usually.

Sometimes you can rig bolted-on arrangements to hold broken cast-iron pieces together, but, frankly, if an old iron stove is valuable enough to deserve repair, I would want to do my best to recast any broken parts that I couldn't repair. But before going to the trouble of making up foundry patterns, I'd check a lot of antique stores for a similar stove. Also, some of the modern stove works have inventories of old parts. If the maker of your stove is still in existence, I'd give him a try. (Just don't bother the Glenwood people if, like us, you have one of their old stoves; they don't make wood stoves any more and have no parts, or so they told us when we asked. Actually, maybe we should nag them and other former woodburner makers. Maybe they'd go back into production.)

## Uncovering the Flaws

Many stoves will be missing parts, and just what to look for we can only discuss in generalities. Of course, you want all doors, lids and such to be there. Don't worry about inside grates that may be burned out in a stove that was used for coal. For wood use you can support logs on brick or adapt a log rest, andiron or the like to the stove. Do be sure the body is sound, though, particularly the floor and sides of the firebox, which are the most likely parts of the body to be burned out or cracked. If the stove has grates holding the fire well above the bottom, a rusted-out floor or cracked side can be fitted with a sheet metal liner; add a layer of sand to the bottom and the liner will serve to catch your ashes. Indeed, you could line an entire iron stove with sheet metal, so long as joints are sparkproof, but since most of the old models were cast in intricate bumps and curves, it would be quite a job. And would need replacing in a couple of years.

Holes in the body are a bad sign, even if the stove has a shiny new coat of stove black. In fact, they are an even worse sign if it has been prettied up; the seller may be camouflaging a broken-down stove. Holes can be puttied up and painted over, masking them till your second or third fire. Were I you, I'd feel free to go over the entire thing with a hammer. But go gently; there's no use knocking holes in weak but still serviceable iron, particularly if the damage you do buys you a bum stove. Listen for an unusual thunking sound rather than the dull ring of iron. You may have found a patch. If it comes loose and is bigger around than your little fingernail, I'd forget the stove. You can keep plugging up holes with stove cement or a blacksmith can hammer in a button of wrought iron if the casting is otherwise sound, but all will work loose in time.

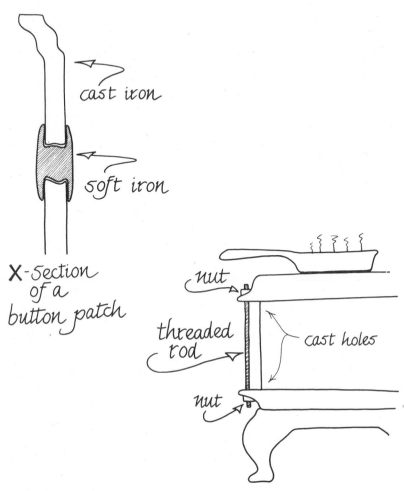

cast iron

soft iron

X-section
of a
button patch

nut

threaded
rod

cast holes

nut

## Missing Parts

Look especially well at the hinges of all doors or opening tops. A broken hinge or rusted-shut draft slide may mean a ruined stove. Nearly all iron stoves were made up of several distinct castings that are held together by stove bolts. The bolts are of soft iron and can be chiseled off or drilled out easily. If they are rusty or missing, don't worry if you are handy with tools. Do be sure that all bolt holes are intact, or at least capable of accepting a bolt. Many stoves are held together by lengths of rod, threaded at each end; the ends go through holes cast into the two parts and nuts are tightened down to hold the stove together. These too can be replaced.

Such removable items as handles, isinglass peep windows, bright metal trim or sliding draft controls are often missing, and many can be replaced. Just be sure the bolt holes are sound. Some will be threaded—threads molded into the iron. Cast iron is devilishly hard stuff and you'll never be able to rethread a hole in it. I suppose there are diamond bits that will bore into it, but I don't have any. Unless you do, I'd not buy an old stove with the idea of boring holes in the iron to repair or replace a missing or broken part.

### Living With the Damage

Quite often, if you are after functional rather than aesthetic characteristics, you can have replacement parts made up of wrought iron or sheet steel. This can be drilled. I know one or two blacksmiths who can take a slab of soft iron and hammer out a presentable facsimile of a missing stove leg or whatever. A welding shop could do the same from angle iron. I've seen a one-legged stove functioning perfectly well resting on three piers of brick and the good leg. A little laundry stove that we almost bought some time back had the door made of old pink Russian stove iron—a good job and if anything making the stove more interesting than if the original cast-iron door was still on.

You may find old stoves with warped plates and open seams between castings. (I'm not referring to cracks in a plate.) These should lower the price a good deal unless the stove is a real oldie. But warping comes from overfiring, usually with coal, and a moderate skew to the fireside of a kitchen range or the back of an old heating stove you'll be using with wood is not always a reason for serious concern. I think I'd plan to install a firebrick liner or find cast-iron liners if you can, and treat the old stove gently. Fire up slowly and don't ever try to get it to radiate much more than a gentle warmth. Cracks between seams can be filled with furnace cement if they are thin—an eighth of an inch at the most.

I've never bought a used soapstone stove, but I can guarantee that any real oldies you find will have cracks in the stone. Don't worry as long as the stove appears sound and the metal frame is holding together. Soapstone just naturally cracks in time. Also, don't worry about eaten-out firebrick in an old stove. Firebrick was for coal-burning models and you don't need

a
one-legged
stove

it with wood. If it has burnt through to the metal, though, plan to put in a patch. File grooves in the old brick and apply fireclay in successive layers as thick as you can make them, till there is a good inch or two between fire and the stove's shell. Don't knock it with a stick till it's fired up good and hard. It will be powder at first. But heat turns it to rock.

So, in sum, look over any old stove carefully. Your common sense will tell you what should be there; just make sure it all is. Go through the mental exercise of firing it up, removing ash, cooking on it if there's a range top; mentally move it into your place, attach a stovepipe and all. If something that should be there isn't, the exercise should turn it up.

## Make Your Own Stove

There is a lot of interest in homemade stoves these days and with the price the stores are getting for cast-iron models of any size, building your own makes sense. You surely aren't going to do your own iron casting, and unless you've a professional welding outfit and the skill to use it, boiler plate is pretty much out of the question. (Though plenty of small welding shops will cut and weld a stove for a pretty reasonable fee if you want to try your own hand at stove design.) I suppose you could build yourself a brick stove on the Rumford design and install ready-made glass doors, though few will try.

The typical handyperson (Louise being a steadier hand with a saw than I) is pretty much limited to sheet metal. Cutting metal of any thickness with common home power tools is tediously time-consuming, and you run through a lot of saw blades. Plus bending and punching thick metal to make rivet joints takes heavy machine shop equipment. Making up a sheet metal stove from scratch is another job that a few may try, but only a few. Fortunately, industry provides the raw material for good stoves that you can get for just a few dollars—oil drums.

a museum
dumb

There are three main sizes: the 55-gallon drum is the most common and easy to find and will heat a really big garage, workshop or cellar. A 30-gallon drum is harder to find, but is a size to fit in with the scale of a home. It also heats as much area as one of the small Scandinavian stoves, though it can't be made as airtight and fuel-efficient by any method I know. And the 15-gallon drum makes up into a nice one-room heater. Whatever the size, it's an "Alaskan" stove. The forty-ninth state is still heated in large part by drum stoves.

Pick up your drum from almost any auto service station or factory, or get one from the drum cleaning service you'll find in any industrial area. Get a fixed-head drum with both ends crimped on. The kind with head held on by a thin strap could fall apart in time. The 30-gallon drum we have cost two dollars uncleaned, and I probably could have gotten it for less if I'd haggled. You can make up your own design. Set the drum up on one end or have it horizontal. With some cutting and welding you can set one, two, or more drums in a stack, connected by heavy steel tubing. Put the stovepipe in at the back of the top drum and you've a modern version of the old sheet metal "dumb stoves" you see in museums. (Some of these had half-a-dozen or more tubes leading from one chamber to another.) The more metal you have heated by fire and smoke, the more efficient will be the stove.

### Drum Stove Kits

More and more firms, large and small, are offering drum stove kits consisting of a cast-iron or heavy metal door, legs and smoke boot or pipe attachment—all you need but the drum. Some include grates, which aren't really necessary in a drum stove. And, sad to say, some are poorly designed and made, probably by someone who jumped on the popularity of wood heat a bit faster than he should have. We ordered one kit that promised to let you see the fire. It consisted of a Pyrex pie plate and assorted flat metal stampings plus an indecipherable set of instructions. That was some time ago, and there are good kits available now, including ones with see-through doors. Most will cost under $50. Some of the stove foundries offer cast-iron kits, but we got a steel-fabricated model from Markade-Winnwood— address in the source list. It has a door that seals nearly airtight-shut, sturdy legs and a smoke boot with a built-in damper and is well worth the cost. In steady use, a drum will burn out in two years or so, then you just remove the kit parts and put them on a new drum. You may have to cut frozen-on stove bolts, but the rest of a good kit should last a good many years.

There are little sheet metal stoves appearing now in the hardware stores that sell for just a few dollars more than the materials for a drum stove. But in the ones I've seen, the metal is thinner than in oil drums, and the doors and smoke boots tinnier than in any of the better kits. I'd say, if you want a good stove cheap, a drum and kit is your best bet. Let's build one.

## Step-by-Step

The photo series shows the basic steps. Following kit directions, I cut door and smoke holes and drilled holes for various bolts. Then legs went on. Wanting a water-heating capability, I worked up a coil arrangement of copper tubing bent so the main run of tubing rests on the bottom of the drum, the ends exiting from holes in the back. Most drums suitable for stoves have a pair of holes, a bung and an air vent. If not, drill them where you wish. Our coil was designed to be pushed in through the narrow door opening. You could make up a coil that goes all around the inside of the drum and sort of screw it into the drum, I guess, but that's more work than I wanted to get into. As is, the coil is buried in the sand that covers the stove bottom up to the level of the door. A little pump, powered by a small electric drill and obtainable for under $10 in any hardware store, will keep water running from the storage tank when the stove is fired up and installed any place with 110-volt electricity. Next fall when it's time to cut the following year's wood supply, my woodchopping partners and I will take the stove out to the handmade shack that keeps us more or less dry at night. We'll try a gravity system, and see if heated water will naturally circulate as it does in the old, but better designed units used with coils in wood cookstoves. This may not work; it just might go to steam and blow us up without a mechanical circulator—we'll be careful at first. However, a small twelve-volt DC pump from one of the mailorder surplus supply houses should work off the batteries of one of the trucks.

Into the stovepipe I put the drum oven, and installed the

updraft system employing two dampers in a T located on the smoke boot that we illustrate on page 184. This way, we can avoid the roaring blaze that a too-strong draft can cause in a nonairtight stove, plus by using the damper in the boot together with the two dampers in the T, we can regulate the amount of heat getting to the oven; close the bottom damper in the pipe and open the other two and you've got biscuit-cooking heat in a minute. To cool the oven, put in a cold brick or open the bottom damper and regulate the other two depending on the fire's heat.

Total cost was well under $100, $35 for the kit, $2 for the barrel, $3 for the little pot rest that goes on top of the barrel, making a flat cooking surface, around $25 for the drum oven, another $2 or $3 for the water storage tank, about $10 for the pump, which has many other uses, a few more for the tubing and fittings, and we already had the electric drill and soldering equipment. Hard to beat for a complete space heating, water heating, top and oven cooking unit, wouldn't you say?

One caution, though. Industrial steel drums are used to transport some pretty noxious substances. Ours held one of those herbicides they use to keep brush down along roadsides. Not wanting any living thing to be exposed to the possibility of breathing the stuff, I put the stove in the pickup, drove to the woodlot, fired it up and stayed downwind till all the paint had burned off the outside and the smoke contained not a whiff of chemical odor. Hopefully, I didn't gas any owls. And that's where the stove is now, in the sugar shack out at the woodlot, with a rust-preventing coat of stove black over all. When we go woodcutting come fall and during the spring maple sugaring weeks, it'll be good to have a warm place to sleep, hot water to wash up in and oven-baked biscuits to go along with the ham and eggs in the morning. The only disadvantage of this design compared to the little potbelly it replaced is the small flat-cooking surface, capable of holding just one pot or skillet at a time. (Without the pot rest on this design, there'd be no skillet cooking at all.) The little drum oven more than makes up for it, though. But if you don't want to bother with the accessories and will be cooking a lot, build your stove up on end and exit the pipe out the back, not the top. The top will have a slight bump to it, but a good cast-iron skillet and your Dutch ovens won't mind.

Assembling a stove from a kit and an old drum is not difficult. First, mark the outlines of the kit parts for cutting and the locations of holes for drilling. For cutting the smoke hole and loading door, use a saber saw (with a blade designed for cutting metal) for the straight runs and a cold chisel for the

corners and curves. The final assembly step is to bolt the kit parts to the stove. Haul the stove outdoors—to our woodlot, in this case—and burn out the drum to remove paint and residues of whatever the drum originally contained. In less than an hour of burning "squaw wood," dry twigs and fallen limbs that burn hot and fast, I had the noxious smell fired away. Back at the house, I used a wire brush and electric drill to scour off the remains of the drum's paint, then coated my new stove with stove black. Though a messy job, a good application

## Central Wood Heat

I guess the ultimate form of wood heat is wood-fired central heating. That is, forced or gravity hot air, circulating hot water or steam systems. Till the oil burner was put in, the house back on the farm was "centrally" heated with wood. In the cellar was a huge old sheet iron and firebrick woodburning furnace that exhausted through stovepipe running half the length of the cellar and into a flue built a yard out from the house (by folks who'd lost one house to a flue fire and weren't about to lose another). The whole furnace was surrounded by a galvanized-tin skin forming an airspace a few inches deep all around. The top of the skin opened into a big wrought-iron grate that was still in the floor last time I looked. A fan rigged in back pushed air through a baffle system so it circulated all around the hot iron before pouring up into the house. Gratings in the first floor ceiling let warm air into the upstairs, and most of the house stayed warm enough, though the folks who last used the furnace claim you

annually will protect the stove from rust. I added a couple of special accessories to my drum stove, a stovepipe oven that is a commonly available item, and a water-heating system that is of my own construction.

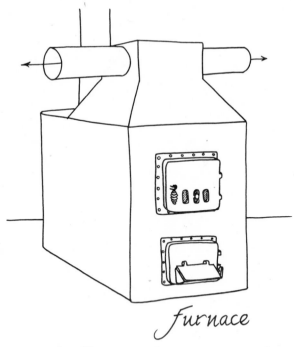

*Furnace*

could toast marshmallows over the iron grating when she was really fired up.

Plenty of older homes have old coal burners that have been converted to oil, simply by putting an oil burner in and closing down the draft openings. Few of these have a large enough fuel box or the properly sized venting system to use wood and I'd not try reconverting a coal burner to something it never was. Such a furnace might adapt to burning charcoal, though, particularly if it is an old model designed to use bulky, clinker-producing bituminous coal. Few folks will have the necessary amount of charcoal available, cheap, but if you do, get ahold of the oldest employee at your local coal dealer (if he's still in business) and ask him to come out and see what can be done with the furnace.

Despite the past availability of cheap petroleum-based home heating fuels, the wood-fired central furnace never completely gave up the ghost. The Riteway firm in the United States, several Canadian companies and a handful of European manufacturers kept turning out small numbers, each handcrafted in large part, so quite expensive. We considered installing a wood burner in the forced hot air system back on our homestead some years ago; the price for the furnace alone was about $7,000.

'Central' heating

Needless to say, we stayed with our conventional wood stoves. With increased demand though, sales volume is going up and prices are becoming more realistic. Also, any number of innovative new designs are coming on the market.

## Modifying an Existing Furnace

Possibly the most practical units and certainly the cheapest are a new development, large wood stoves that are designed to augment your existing oil or gas-fired furnace. Each maker has his own proprietary system for tying into the existing furnace, but the principle is the same in each. Your current automatic system must be rewired (according to building code by a licensed heating/plumbing contractor) and either a woodburning hot air or hot water stove hooked up to the furnace. The hot air unit is a wood stove enclosed in a sheet metal box that has a cold air return, which sucks in house air, and a hot air outflow, which puts heated air into the furnace plenum to be circulated by the furnace fan. The hot water unit has a set of coils in the firebox. Furnace water is pumped through the stove and back to the circulating system to be pumped up into the radiators. Each unit has a sensor that is installed in the furnace proper; it turns on the furnace water or air circulator as indicated by the house thermostat. Then when the wood fire dies down another gadget brings in the conventional burner. I've not seen such an arrangement personally, but I'm sure one of these units could be easily integrated with an electric home heating system too.

These units will cost something over $500, which is what the larger conventional stoves will run. I'm no heating contractor, and I'd want to have one do a careful appraisal of any system that ties into the existing furnace before I bought. Also, I'd want a really careful evaluation of the flue situation. None of the sales pieces for stoves intended to augment an existing furnace mention how the smoke is to be exhausted. Presumably they expect you to tie into the furnace stack. Well, as noted elsewhere, that can cause problems. Of course, the oil or gas burner is presumably not going when the wood stove is fully fired, so if the stove is vented into the chimney at a spot at least two feet from the fireplace smoke hole, there shouldn't be continual interference or blow-back between furnace and stove drafts. However, what happens after the wood fire gets cold enough that the burner turns on? There will still be coals, smoke, open draft holes in the stove (I've seen none that shut off air flow automatically when the burner goes on). So, the stove will continue using a good proportion of the flue's drawing capacity while the burner is on. The result could be incomplete or too slow evacuation of oil or gas stack gases, producing an inefficient, dirty flame that wastes fuel and soots up the burner and fire chamber, further reducing efficiency. So good-sounding though it may be, I'd suggest that you take on a woodburning supplement to an existing system with considerable care.

### The Original Woodburning Furnace

More time-tested are the original wood central furnaces. They are two-stage burners with a lower chamber for primary, an upper chamber for secondary combustion. Fuel efficiency is greater than with most space-heating stoves or any fireplace, and creosote buildup is lessened. Heat from both chambers is passed through a heat exchanger to a hot air plenum or a water or steam boiler. They all have automatic draft controls, some electric, others (simpler but more foolproof) operate on the bimetal strip principle used in circulators.

Chances are that you'll have some difficulty getting any kind of wood unit approved as the major furnace in a home located in any urban or suburban area. So be sure to check the building code and get needed approvals before sending off for a furnace. In traditional, usually rural, woodburning areas in the Northeast, wood furnaces are okay by most codes. However, even in our neck of the woods, there are code-compliance problems with the most versatile of the woodburning central furnaces, those that also use another fuel, usually oil or gas, as a back-up for the woodburner. The units are simply too new to the market—or to a widespread market—the concept has been around for years—to be built into most building codes. You may have to build a new Class A flue for the woodburning part of the stove, and there are few licensed heating contractors who know much about the units.

### Multiple-Fuel Furnaces

Still, the multiple-fuel furnaces offer a dandy convenience, particularly if you plan to be away from home at all during freezing weather; with wood heat only, you'd have to drain water pipes and remove anything that might break or be harmed by freezing such as all the food put up in glass jars (though some wood-burning folks arrange little electric heaters strategically in the house and set them to keep things just tepid while they are away).

You can attach a woodburning central furnace to any standard air, water or steam system, and unless you know more about modern heating plants than I do, you'd best have the installation done by a heating contractor, even if the code doesn't force you to. One thing he won't be able to do, though, and that you should bear in mind before opting for a big woodburner, is to keep it fed and remove the ashes. Remember, the central heater is pumping warmth into every room in the house, and unless the place is really well-insulated, that takes a lot of wood. The only house I know well that was heated exclusively by a modern woodburning central heater just happens to be the oldest place in town, and pretty drafty at that. The owners had to stoke it full at bedtime and get up good and early to add more

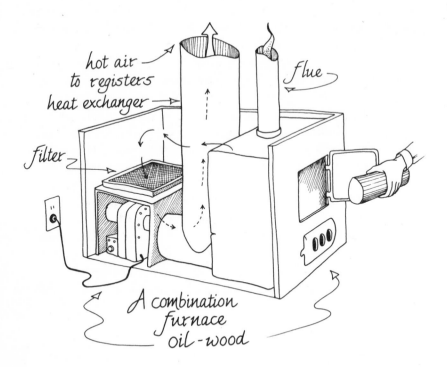

hot air —
to registers
heat exchanger →

flue

filter

A combination
furnace
oil-wood

wood to keep the water in the wash basin from freezing. And they celebrated reaching senior citizenship by installing an oil burner. However, manufacturers of more modern designs have reduced intervals between feedings to about ten hours, or so they claim. Plus the furnaces will accept really big logs, four-foot cordwood, some of them. This greatly reduces cutting and stacking time too, needless to say.

The manufacturer and seller will be able to give you each model's heating capacity and fuel requirements. Just remember, in any kind of central heating, you are putting warmth into the entire house, not just a few selected warm spots as with stoves. More area, walls and floors are kept warm all the time, so you will need considerably more wood than with a conventional stove or two. I don't know anyone with a central-heating wood furnace that doesn't live on a good-sized country place, in the middle of a really large woodlot.

And then there are the ashes. The central heater will be in the cellar likely as not, and at season's end you will have four to five hundred pounds of ashes on your hands. Well-combusted wood ash is light and fluffy stuff when fresh, and a pound takes up the better part of a cubic foot or more. Unless you wet it down in a lye tub you'll have to store or haul off a great deal of ash. Out on the farm where appearances didn't matter, we just threw excess ash up on the snow covering the garden. In town where that wouldn't be neighborly, the extra is stored in a brick ash pit that came with the house. Without it, we'd have to put the ash in sacks. If we had a central heater instead of the stoves, I imagine the cellar would be pretty well filled with ash sacks come spring.

Still, if you have plenty of wood and the ambition and equipment to get it into the woodshed, a woodburning central-heating unit can get you as free of the oil companies as you wish to be—and in a house that's heated in the most genuinely modern way I can think of—an up-to-date central-heating system fired by the old-fashioned fuel of our petroleum-short future. And you can still have a wood range and all the fireplaces and heating stoves you want.

## Installing a Stove

Well, by now you've decided just what size and style of stove you want, and likely have it sitting in the center of your living room. Before installing it, consider first the cardinal rules of safe stove installation, which I must admit are honored in the breach more often than they should be by many folks, including us before we really boned up on wood heat. I strongly advise you to follow them to the letter, even if they are more stringent than the rules in your local zoning codes.

### The Cardinal Safety Rule

In essence the rules boil down to one sentence: no wood stove, stovepipe or flue should heat combustible material of any kind so hot that you can't rest the palm of your hand on it indefinitely. Now, this runs counter to a lot of old-time wisdom you may pick up, such as the pre-turn-of-the-century practice of using the flue to support main floor timbers in the house. Well, even a sound chimney can heat up with a too-hot fire and can bring a dry, old floor beam to the point of combustion. Today, a minimum of a two-inch gap or firebreak is built in between wood house parts and flues. If your old chimney doesn't measure up, replace it, reline it or keep your fires low and stack gases cool.

Lots of old stoves sit on high legs, and in the old days they were placed right on the floor, or on a stove board of thin wood covered by a sheet of metal. Now, one of the superinsulated circulating heaters or a cookstove where the firebox is located a good yard above the floor is safe on floors or a stove board. No other stove is. Check the photos of the several stoves in the book. All but the cookstoves rest on some sort of stone or ceramic base, and these plus legs give enough height to satisfy the quantified version of the main rule:

*There must be at least 18 inches of space between a radiating surface and any combustible material.* Period! That applies to stovepipe as well as the stove.

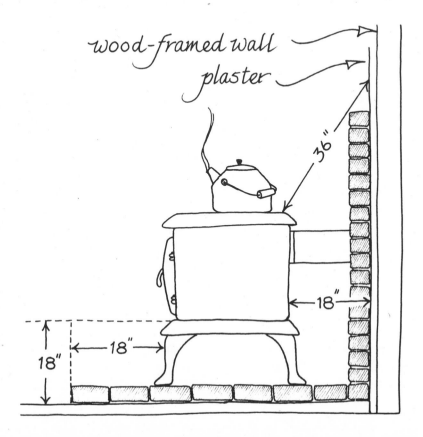

Increase that distance to a good yard between any combustible surface and the top, back and sides of any radiating stove. Some insulated circulators can go anywhere, and, of course, you can back a conventional stove right up to a bricked-up fireplace with a good ceramic hearth—so long as the mantle, wood framing around the fireplace and any combustible wall is a yard from any radiating surface. The hearth must be of a non-combustible surface, but since heat rises, you can get by with just a layer or two of bricks over a wood floor. Just make sure that the combination of stove leg height plus height of the hearth material equals that 18-inch minimum above the wood! The hearth should also

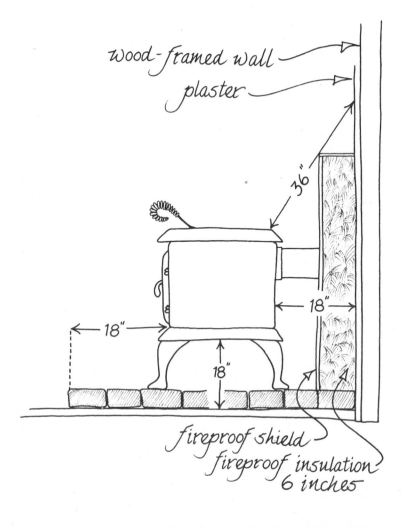

wood-framed wall

plaster

36"

18"

18"

18"

fireproof shield

fireproof insulation

6 inches

extend at least 18 inches out around the stove on all sides—
sparks do fly.

One respected fire safety organization says you can have
your stove on little four-inch legs if there's a sheet of 24-gauge or
thicker metal underneath for 18 inches all around the stove. I
don't go along. Sheet metal can heat up and transfer that heat to
a wooden floor beneath. Louise and I stick with the 18 inches in
combination of leg length and cinder block, bricks, sand or
whatever between the bottom of any stove and a wooden floor.

The same clearances apply for brick-over-frame construction
used to make walls noncombustible; however, the stove is still
18 inches from the framing behind the brick and plaster. If
building a masonry backdrop is inconvenient, install a heat
shield, a metal or asbestos/metal sheet an inch or more out from
the combustible material in the wall.

With the barrier (and if a permanent installation, a layer of
noncombustible insulation such as fiberglass between wall and

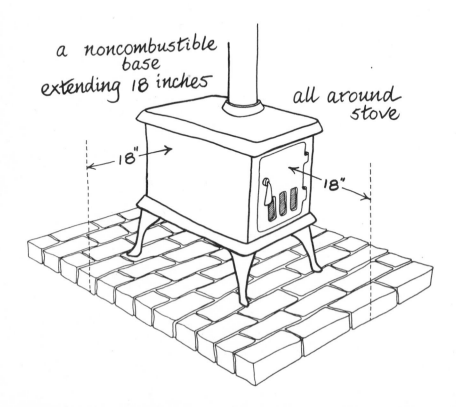

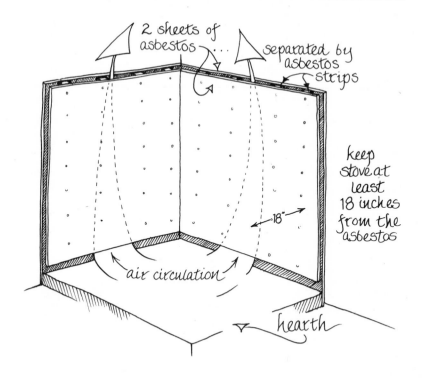

2 sheets of asbestos ... separated by asbestos strips

keep stove at least 18 inches from the asbestos

18"

air circulation

hearth

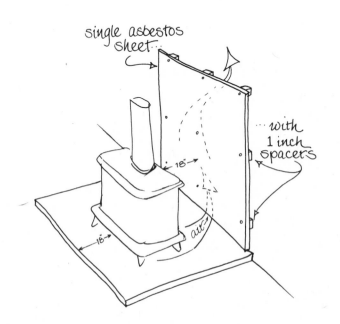

single asbestos sheet

with 1 inch spacers

18"

18"

air

board) the stove can be placed 18 inches from the wall. If you've any uncertainty about the safety of your own installation, and we haven't covered it here, get *Heat Producing Appliances Clearances* (NFPA no. 89m-1971) and *Using Coal and Wood Stoves Safely* (NFPA no. HS-8-1974) for a few dollars from the National Fire Protection Association, 470 Atlantic Ave., Boston, MA 02210. And follow their advice. The publications are illustrated with photos of house fires caused by improperly placed wood heaters and captions give the numbers of people who died if you have any doubts.

When the main heating stove in the house was installed sometime back, we had an experienced mason erect a new flue for it. But the masonry hearth and backwall we built ourselves, as you can see. I bought a standard "cube" containing 544 bricks, selecting the kind with holes in them because they are

lighter, easier to split, and cheaper. Splitting the bricks is done using a mason's chisel and hammer. I found that a single sharp crack with the hammer splits best and cleanest. We chose a corner location and a "sunburst" brick pattern for the hearth. After stapling a layer of asbestos paper fireproofing to the floor, I used a length of string as a compass in marking arcs for the courses and as a centerline in aligning the bricks. The bricks were positioned, and I dusted sand into the joints, filling out the final half-inch or more with mortar. A mason

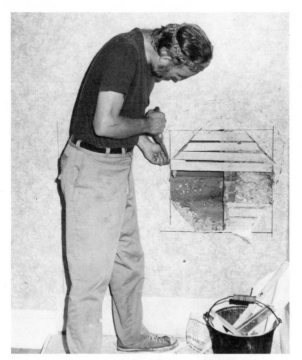

doubtless would not do it that way, but I'm no mason. Next, I outlined the firestop box and cut away plaster, lath and wall framing as necessary for it. The firestop box I constructed as explained in the chapter on flues. After lining the wall with the asbestos paper, I laid up the bricks, cheating just a little by scribing lines on the wall so I'd know where each brick was to go (and that it was level and plumb without constantly checking a level). Metal wall ties were nailed to the studs and mortared between bricks. I finished up the firestop box

When you are planning the location of your stovepipe, remember, it heats up hottest at bends, both the elbow, if any, at the back of the stove where the pipe changes from horizontal to vertical, and again at any other directional changes along the pipe run to the flue. Hot gases pour out of the stove and collide with the elbow, which forces them to change direction. Often the elbows will be the hottest part of your fire and may need special attention if they approach a combustible surface. Make the "never too hot to touch test" if in doubt.

### The Damper

Many stoves come equipped with their own damper—that round plate a bit smaller than the stovepipe that revolves on a pin to regulate flow of air and smoke out of the stove and into the flue. With a stove lacking its own, you'll have to put one into the stovepipe. Most commercial dampers come with a spring-loaded pin. You can push in on the coil-wired handle and the pin will

assembly after the brickwork was finished, by loosely wrapping a six-inch stovepipe with asbestos paper, slipping it inside a seven-inch pipe, wrapping the seven-inch pipe with asbestos paper and slipping it into the eight-inch pipe in the firestop box.

come out of the guides in the plate. Then punch right-sized holes in opposite sides of the pipe (horizontally usually, though any way will work), put the pointed end of the pin in one hole, put the plate into the pipe, run the pin through guides in the plate, and out the other hole in the pipe. A bit of pressure and a twist on the handle to join the parts and you've a working damper.

Where you install the damper on the pipe is up to you—wherever it is handiest. You'll be using it a lot, opening it full to warm up the flue to get a good smoke-clearing draft when starting a fire and just before you add wood, and dozens of times a day as you use this rear air flow control in conjunction with the front draft controls to regulate the fire. So place it in the pipe where it will be handy when you are kneeling down in front of the stove with a load of wood in one arm.

Many stove makers and sellers will try to tell you that you don't need a smoke damper with their airtight designs. Some Scandinavian stoves even come complete with enough pipe (sans damper) to rig the stoves into a flue. But, I don't buy the idea, perhaps in part because when I was too ignorant to put a

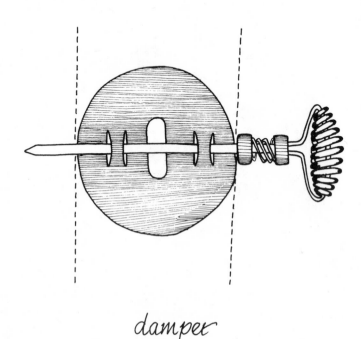

damper

damper into the pipe back at the homestead I nearly burned the place down. Well, that was with a leaky Franklin. With an airtight you can regulate the fire by use of the draft control—but that's all the control you have. On cool days or with especially dry wood and in any other number of circumstances, you'd have to close the door and shut the draft down to keep from roasting yourself out of the place. Well, this will produce a super-creosoty fire. With a damper you can control draft at the smoke end and keep a small and healthy blaze in any stove. A damper doubles your control over your fire, and I wouldn't install any stove without one except for models such as central heat furnaces that are rigged for fully automatic operation.

If you've an imported stove with metric-sized smoke hole and pipe, you may have a hard time finding a properly fitting damper. You can buy an inch-sized damper that's a bit smaller than your pipe and have a metal shop weld on a proper-sized plate. Or you can fit a metric-to-inch pipe adapter to the smoke hole and run a larger size standard pipe and damper. Indeed, this is what I, among others, recommend to anyone with an airtight stove that comes with a little four-inch smoke hole (common among the imports.) Move up to a six-inch pipe with damper and your stove will be able to breathe easier, will be able to run with a small hot fire for much of the day, will heat the flue faster when starting cold and accumulate much less creosote all along.

### Which Way to Run the Pipe?

Speaking of stovepipe and creosote, there's some controversy over how the pipe should be run, whether the ribbed or crimped end (that fits inside the uncrimped end of the next section) should aim away from the stove or toward it. Aimed away, the smoke has a smooth flow to the flue, but if pyroligneous acid forms in a poorly mated pipe, it can leak out. If the crimp aims down, smoke can leak out of the pipe, but creosote will stay inside the system. Here again we have the old-time wisdom of past generations competing with the new-found wisdom of the airtight stove experts. Old stoves are all designed to mate with the open, uncrimped end of pipe. Crimps head up and away. They were also designed to burn hot so wood volatiles would go into heat, not creosote; the longer the run of stove-

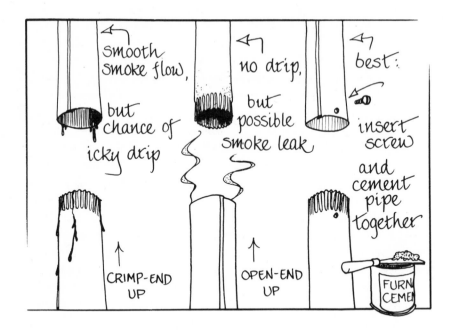

smooth smoke flow, but chance of icky drip

CRIMP-END UP

no drip, but possible smoke leak

OPEN-END UP

best: insert screw and cement pipe together

FURN CEME

pipe, the hotter they would be run. If any creosote did condense out, it would be in the flue and ideally, the flue was kept hot enough that most of the junk that did escape the firebox would go right on out.

Now with the airtights lacking dampers, with their often tiny stovepipes and puny air supply, new stove owners are finding sticky black stuff oozing out of stovepipes everywhere. When we first began running our own Combi, we had a tremendous liquid creosote buildup, too—it ran into the house and down the brick-work, down the clapboards on the outside too. It is surprisingly sneaky stuff; it can even flow through brick joints. The problem was a too loose series of pipe connections. Cold air was leaking in and chilling the smoke when we burned the stove in the closed-up configuration. I closed up the leaks by hammering the pipe together more securely and by packing fireclay into the places where the match was less than perfect—typically at the seams in the pipe. You might want to use furnace cement; just pack it into the crimps and whack the pipes together. That makes for a more permanent joint than we wanted, however.

Now, this worked; our pipe happens to run horizontally into the flue connector and even if it had been joined crimped end aiming toward the stove, we'd have had leaks—and probably more in fact, as that rig does cause smoke eddies inside the pipe, slowing smoke flow. So in my experience, the old-time way of joining pipe, crimps aiming up or away from the stove is still best. If you've got creosote leaks, I'd say it's a symptom of an installation or firing problem that you are better off knowing about than masking by reversing stovepipe (and just increasing the problem by disturbing smoke flow).

Presence of liquid creosote at any point in your wood heat system means you have air leaks, are burning green wood and/or don't have fires hot enough to create a good fuel-efficient degree of combustion, and produce a hot enough draft and smooth smoke flow. So rather than reverse the stove pipe, check the system. Are there any leaks? You should have a perfect, airtight seal between door of the stove and flue top. Is your run of pipe too long? Perhaps you've put in a flue robber, a heat exchanger that helps heat the house but overcools the smoke. Or, as grumbled about earlier, you may be running the stove too cool. I know that this flies counter to much newly conventional wisdom, but, as in many other aspects of wood heat, I am finding many of the new ideas proving inferior to what our great-grandparents proved out over years of practical experience.

### Two Easy and Safe Installations

Probably the easiest and safest way to install a stove is to put it into or in front of an existing, safe and soundly mortared fireplace. Several traditional Franklin-style stoves and newly designed fireplace inserts are expressly designed with front-operated dampers for this application. If the size of the hearth is big enough to provide the 18-inch floor space required all around, and wooden mantles, et cetera are at least a yard above and to the sides of any part of the stove that could radiate heat, just put her in, connect the damper and run it into the flue. Stuff fiberglass insulation or stove putty around to seal it up and you're ready to fire up. If clearances are insufficient, and/or if you want the stove well out into the room so it will radiate most efficiently, you can lay a fireproof foundation and run your pipe back and up into the flue.

Many older homes have had fireplaces bricked up and stove holes punched into the covers. There are several firms market-

*fireplace insert –
a Franklin-type stove*

front
damper

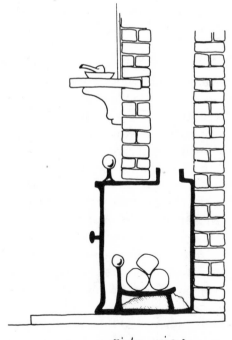

*side view*

ing fireplace covers now too—you just put one over the fireplace opening and clamp it on at the lintel. Back in the old days, such an installation was safe enough. Older stoves were free-burning and used plenty of draft to pull hot, well-scrubbed smoke up the big old flues. But I have serious questions about venting a modern airtight into an old bricked-up fireplace or installing one with a new-design steel fireplace cover. We've gone on at length about the creosote production of cool, low-draft heating fires. Well, here you are hooking an airtight designed for a four- to eight-inch stovepipe and flue into a great cavernous fireplace with a foot-square or larger flue designed for a blazing grate fire. The fireplace-cover makers haven't proven (at least not to me) that their covers are perfectly airtight, especially around the floor. So there's a chance that cool floor air will be leaking in around a cover, and the stove draft thus may be too low to pull smoke out of the flue quickly. Cold outside air can fall down into the firebox and chill the already relatively cool stove smoke. It just seems to me that this could leave almost all the smoke's volatiles in your fireplace as creosote, with a resultant danger of fire. Of course, with a modern removable cover, you can take the thing off to check for creosote formation and to clean if necessary. But there's nothing you can do with an old bricked-up fireplace except wait for your house to burn down. My advice is to open any bricked-in fireplace and vent the stovepipe directly into the flue. If it were my house, I'd do a lot of checking too before buying a new fireplace cover for an airtight stove. Personally, I'd be a lot more comfortable spending a few dollars to have a sheet metal shop make an adapter that would connect my stovepipe directly to the throat of the fireplace at the top of the firebox. I'd have to measure the area of the throat (where the damper assembly is located) and stay with a stove needing a stovepipe no larger than the area of the throat. If you've forgotten your high school geometry, multiply the width times the depth of the throat to get its area. Compare that with the area of the pipe, obtained by multiplying the radius of the pipe (half the diameter) times itself, then multiplying this by three. The area of a six-inch pipe is about 27 square inches; a seven-inch pipe, 36 inches; an eight incher, 48 square inches.

By the way, let's touch again on the number of stoves you can vent into one flue. Basically, the combined area of all stove-

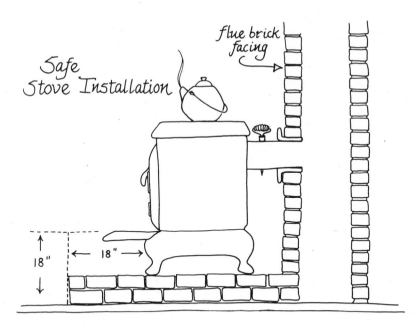

Safe Stove Installation

flue brick facing

18"   ← 18" →

pipes exhausting into the flue should be a bit less than the area of the flue itself. Assuming that you have a fairly typical foot-square flue interior, that means you could install five six-inch pipes, four seven-inchers or three eight-inch pipes. That is, if all are operating all the time. Theoretically, you could pepper the flue with as many stove holes as you can get in without causing it to collapse, so long as only the above numbers were operating at one time. I'll have to repeat once again the admonition that your central heating *should* be on its own flue, and so should each fireplace. We live in a house where a central-heating system has peacefully shared a flue with a wood cookstove and two room heaters for the better part of a century, however.

## Safe Operation

If you've made sure that the flues are safely constructed, then be just as sure the stoves, pipe connectors and your own operating techniques are too. Wait till dark, then put a lit flashlight into your stove and close all doors, dampers, and drafts. Check carefully to see if light is coming out anywhere it shouldn't. Caulk any cracks, say where castings fail to meet, with stove cement. You don't want an errant spark to come out

where it isn't expected. It's a good idea to repeat this check periodically.

Be sure that all stovepipe connections are firmly mated. As mentioned earlier, loose connections can emit sparks, let pyroligneous acid out, and affect the draft. If your pipe runs any distance in the horizontal, wire it securely to the wall or ceiling. As long as the pipe is always running at a slightly upward slope as it moves away from the stove, you are burning open nonairtight fires, and you clean the pipe frequently, a well-secured pipe can be as long as your room is wide. I've seen pictures of old school rooms and churches with pipes 30 feet long and more. Again, be sure the pipe is secured well enough that a fire in it won't send the sections flying. I'd recommend securing each connection with three sheet metal screws and wiring the center of each two-foot section of pipe to the ceiling to be on the safe side.

### Controlling the Draft

Some stove/flue combinations are improperly matched—say a small, loosely constructed stove with a six-inch pipe venting

draft
control

into a warm flue that is 18 inches on a side. Theoretically, the flue could handle 18 stoves of that size, and it can exert quite a strong pull on just one. This is usually the problem when stoves roar or overheat despite all you can do with dampers and draft openings. Restricting the flue size at the chimney top might be feasible. Probably easier is to install a draft control. These are devices that admit room air to the stovepipe to equalize draft.

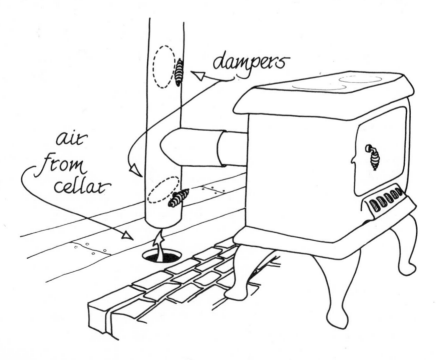

dampers

air
from
cellar

The automatic type that you find on most furnaces has a flap that is pulled open when the draft is more than the stove requires.

You can make your own draft control by installing a T-connector in the pipe and putting a damper in the open end. It can go on the pipe between stove and flue, above or below the stove damper, wherever seems to work best. I've seen several ingenious uses of Ts to tame recalcitrant stoves. One, which will burn out the T in short order, placed it at the boot of the stove, the top of the T aiming vertically. A length of pipe with a damper was aimed down to a hole in the floor, the other, with a damper went on up to the flue. When the stove had more draft than it needed, the lower damper was opened and cellar air went up the chimney, curing the too hot fire but not causing drafts and loss of heat in the room.

## Firing Up the Stove

Finally, let's fire up your new stove. You'll want a handful of small splinters, as dry as you can make them, a half-dozen kindling sticks about an inch thick and three or four medium-sized pieces three to four inches through—"quarter-splits," or your typical firewood-size log split into four parts. Then, you will need a supply of big logs, whole limb pieces and "half-splits" or larger logs, eight inches through and larger. Be sure the wood is warm. Bring it in the day before to let it lose its chill and have time for surface moisture to evaporate. Then open the stove door and damper all the way. This is to let the flue warm. Remember, cold flues will chill the first smoke, sending it down and back into the room; oxygen won't be pulled through the wood, and instead of a well-fired stove, you'll have a smudge pot. Cold flues don't draw, as they say, and they attract creosote, so let yours warm up. If room air won't rise because outside, cold air keeps falling in, light paper strips or a small, dry and hot fire in the stove top or back.

The firing process is similar for any stove with a grate or wood basket, andirons or whatever to hold wood up off the stove floor. This includes the circulators, Franklins and other open stoves, potbellies and their kin. Loosely crumple several sheets of paper between the andirons or into the firebox and lay on a couple of handfuls of splinters. Any which way is the best arrangement as long as there is plenty of space between each.

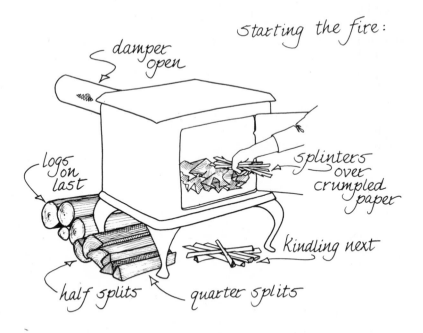

Starting the fire:
damper open
logs on last
splinters over crumpled paper
Kindling next
half splits    quarter splits

Now put on the kindling, in a crisscross pattern in square stoves, a tipi shape in the high railroad types and potbellies, longwise in a series of X's in the long logburners. Some people lay in the larger sticks now; I like to light up, let the paper collapse and be sure the kindling has caught well first. With cold or wet wood and a really cold flue on a really cold day it sometimes takes several starts to get up a good blaze. There's no need to keep pulling out partly scorched wood to relay the paper and kindling.

Once kindling is caught, I put in as many quarter-splits as I can get on in one layer, leaving just a crack between each piece. This is Louise's own magic fire-starting principle; before she figured it out I was taking ages to get a fire going. She's still better at it than I. The narrow opening between logs increases the velocity of hot air flowing up from underneath, in effect "blowing" on the fire to increase oxygen supply at the kindling point of the wood. With split wood there are little splinters sticking out, and splits catch faster than whole logs.

When the first layer of quarter-splits is going well, I put in some larger pieces of split wood, another single layer to fill the

# kindling arrangement in:

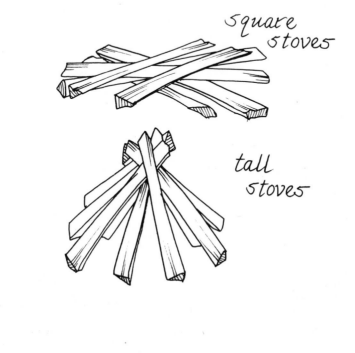

square
stoves

tall
stoves

long stoves

firebox from side to side. In a half hour or so a good bed of coals will be established, and you can fill the stove and close the damper halfway. In perhaps an hour more the wood will be hot and you have yourself a stove fire. Here on in it's between you and your stove to figure out what combination of damper setting and air control will give you the heat you need.

**Maintaining the Fire**

With any open stove such as a Franklin, the best fire is the fireplace-type ash bank that we'll cover in detail in the following chapter. But most closed stoves have two air entries or draft controls in the door or sides. Some have three, the third a skirt at the very bottom of the firebox that pulls out to let a draft roar up through the bottom of the fire, and old model stoves may have a half-dozen scattered around. But in conventional (nonairtight, noncirculator stoves with grates), there will usually be two sets

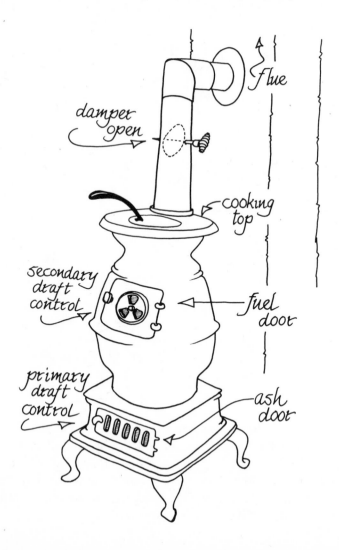

of slides. The bottom one is to admit air for primary combustion, burning of the carbon in the coal bed. The top control admits air for secondary combustion of the gases. How to adjust which control varies with every stove/flue combination, with the type and dryness of your wood and even with the weather. If it's cold outside, the flue will need more warmth to start and maintain a good draft. On a relatively warm day, a roaring blaze for very long is a waste of wood.

We've often found that closing up a stove and opening the bottom draft control all the way makes for the quickest starts. It's air velocity working again. Then when a good coal bed is established, the bottom can often be closed completely so coals burn more slowly, the top opened to one or another degree to

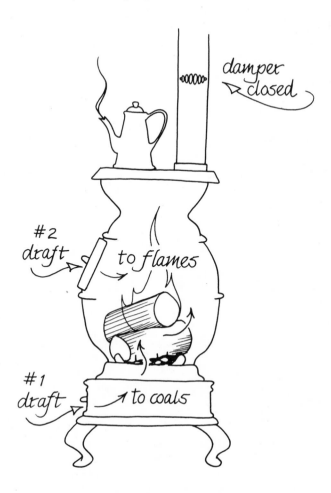

assure maximum combustion of the gases. But the ideal heating fire has the damper completely or almost completely closed, the stove closed up with just enough air entering at each level to maintain the degree of heat you want. The more heat, the more air you let in, the wider the damper opening and the more often you reload with wood.

In almost any stove, reloading a well-started fire is simply a matter of piling in more logs. Unless the wood is wet, in which case you'll often need Louise's magic spacing to get it going, an open-type fire will take wood any way it gets it.

### Firing the Grateless Stoves

Many of the imports lack grates or andirons, and firing them the first time is a different proposition. Open damper and door to let the flue warm as with any new fire. Lacking grates to keep the wood up (in one of the long baffled models) I like to put a short but fat log on crosswise about two inches back into the stove. Crumpled paper goes in back, splinters are scattered on top, and the kindling laid in so the fronts of the sticks rest on the

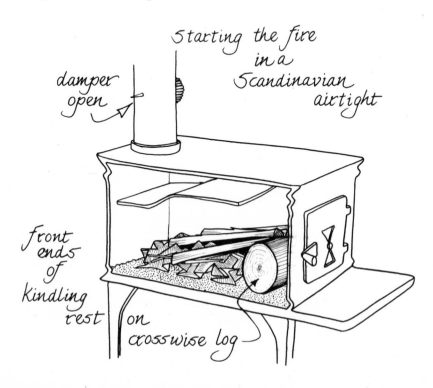

Starting the fire in a Scandinavian airtight

damper open

front ends of kindling rest on crosswise log

crosswise log. That way, plenty of air gets to the wood. When kindling is going, in go the quarter-splits, and then the big logs. After a day or so of burning, you'll have enough ash that you can rake an ash mound replacement for the starter log up to the front of the stove. Then just keep putting in the wood.

With the upright imports, the Scandinavian Combis and a few of the Central European iron and ceramic stoves with or without grates, you can use the tipi arrangement of paper, kindling and logs to get the fire started. In time the tipi will collapse into coals and you can continue adding wood in tipi fashion or as it happens to fall. These high cylindrical or square-topped stoves have a firebox no more than a foot square and if you try to set big logs in on their ends they will often snuff out the fire. You'll have to split virtually all logs more than six inches through to keep a fire—something to think about before taking one on. Folks I know who do use this type of stove—the tall, slim design goes beautifully in many rooms—find that the best fuel is slabwood, the thin pieces cut off to square logs up for cutting into lumber at the sawmill. More on slabs when we get to "Getting In the Wood."

Pajama time for the kids means we can bank the fire and close the big stove for the night. With a full load of logs and the damper almost fully closed, the stove will keep the house comfortably warm for sleeping all night long.

With stoves such as our Combi, which offers the open fire of a Franklin with a well-sealed door of the other super-efficient airtights, you have a choice of fire design. A Norwegian friend starts her fires in the tipi form and keeps them that way. The coals and ashes are maintained in a cone shape in the center of the firebox and logs are laid up against the pile at the sides and pushed down into the ash at back. Dead ashes gradually work their way out to the front where they are removed. Louise and I hark back to ancestors from a bit south of Norway, and we like our cheery open British fire on its andirons or log rests. We couldn't find a set small enough for the Combi, so I got the smallest set of log rests I could find and had a machine shop saw one in half and cut down the legs. They rest on the bottom of the stove, angling down toward the back. During the day we keep the banked fire described in the fireplace section. At night, a half-a-dozen or so logs are stacked in against the bed of ash and coals and the door is closed, damper shut down and draft opened

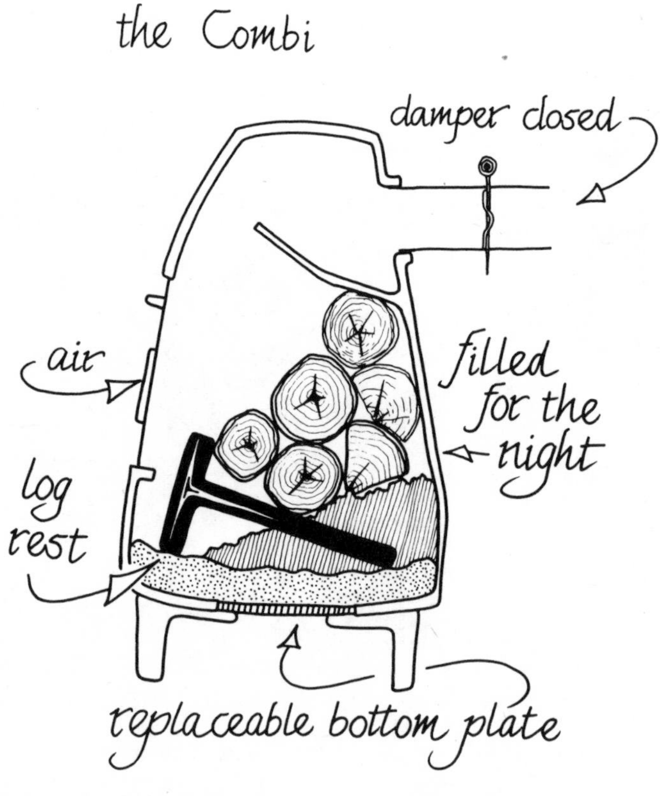

the Combi

damper closed

air

filled for the night

log rest

replaceable bottom plate

depending on the temperature outside and degree of warmth needed. Settings will differ with each installation and you can learn only by trial and error.

Just don't shut any newly filled airtight up completely (or almost completely—all dampers are designed to let some air through). One neighbor closed up his airtight stove the first night he got it, came down the next morning and found a cold room and the stove full of perfectly distilled charcoal! Plus, I'm sure, a brand-new layer of creosote on the stovepipe and flue liner.

### Still More on the Creosote Problem

Sorry to keep harping on creosote buildup in airtight stoves, but it's a problem deserving a good airing. The people I know who have been burning cabinet-style circulators for years acknowledge the creosote problem; a few open the stoves up for a hot burn once a day to incinerate creosote in the pipe and dry what's accumulated in the flue. Others just shrug, let it build up and clean everything out two or more times a year.

With other type airtights, most knowledgeable people I know prefer the cycle we follow with the Combi: open, hot fires of whatever size needed during the day, closed up at night with a full load of good, dry top-quality hardwood. Properly loaded, the big stove will heat the whole house for a good twelve hours in airtight operation. But I do have to clean the flue annually and the flue connector (some four feet of flue tile leading out to our freestanding outside chimney) twice a year. It's part of the price you pay.

Where I get upset is when folks burn the box-style airtights in a closed down mode day and night, packing them full with logs several times a day. This is asking for a crudded up stove (especially if you have one of the lovely Scandinavians with a complex series of baffles in a stove-top heat exchanger) in a month's time. Also asking for a creosote-filled and dangerous stovepipe and flue.

Personal opinion aside, these stoves were not designed to be packed full and closed down; that's how to make charcoal, as our friend did inadvertently, or to turn coal into coke (except that in modern coke ovens the volatiles aren't allowed to accumulate in the stacks, but are retrieved from the stack gases and refined into several valuable hydrocarbons). Nope, to burn any airtight, Scandinavian or other radiating design without a proven secondary combustion function, keep her opened up in front, even if you have to make a special spark screen, and control air flow with a damper in the back. Start a small, fast fire, build up a good coal bed, then add wood in dribbles.

Come night time, at least one manufacturer of Scandinavian stoves recommends you quit loading wood in early evening, but get a good coal bed established, then close the stove up all the way. Heat retained in the mass of the stove plus the coals will keep things warm enough that the pipes won't freeze, and you will have a good bed of coals to begin heating in earnest next day. Now, that's what the manufacturer says.

If you use an airtight in any other way, you are going to get creosote. Not that that is so bad as long as you minimize it and keep it cleaned out. What I do, preferring a warmer house, I guess, than the Scandinavian stove makers, is to select only the most dense hardwood for overnight burning. It is stored inside for several months—and this after a full year curing in the woods. It dries down to a 15 or 20 percent moisture content, so that if burned in an open fire, it would go up far too rapidly. Just before bed time, we scrape out ash at the stove front and build up the fire to a great roaring pitch to get a super-hot ash bed. Then the dry wood is packed in and *let burn hot for a few minutes* to drive off moisture and also drive off and burn much of the volatile oils. (Though wood itself burns at about 600 degrees F., the oils volatilize at a somewhat lower temperature *but* require a much higher heat, around 1400 degrees F, to actually burn. So the brief hot initial fire with most of the heat and gas going to heat up the flue is good preparation for a slow all-night burn.) Then all is closed down but for a small crack in the air inlets, and we stay warm and comparatively creosote-free the night through.

It may seem paradoxical that a slow, relatively cool fire in an airtight stove puts out more useful heat than a hot, free-burning blaze, but it is the truth. You are simply containing the fire and all its components, smoke included, in the stove as long as possible. I admit a bias in favor of antique designs that do this same thing by some of the sophisticated smoke recirculation systems alluded to earlier. Still, the modern day cast-iron airtights, simple in design as they are, will heat you well and efficiently. Just be prepared to cope with that creosote.

### Secondary Combustion Chambers

Now, I obviously advocate the small but hot, flameless and smokeless fire. Indeed, when we get to fireplaces, where you can really inspect a fire as it burns, we'll see that the best heating fire is one with barely any visible flame and minimal smoke. But some folks will want to install the bigger potbelly types and homemade barrel stoves in garages or drafty workshops where the gentle glow of an efficient airtight-style stove fire just won't keep you warm. What you want is a real barn-burner, a fire with a draft that will suck a sheet of paper in and have the front half burnt before the tail end gets through the stove door. Here, you

variations

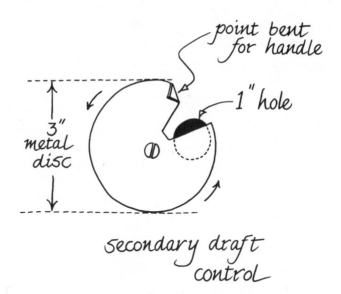

secondary draft control

open everything up wide and keep pitching in logs to get the maximum safe amount of heat radiating out.

This will be a fire with an unavoidable high flame, the sort that potbelly stoves were built for. The high top pretty well contains the flame from the fire low down in the belly. (For high-firing, build your homemade drum stove standing upright too.) And with any stove used for high-flame-type fires, you may want to add on a secondary combustion chamber—an extra chamber where the flame can be contained and a bit of extra oxygen introduced so the gases expelled from the stove by the high flame can burn more completely. Often the stovepipe or flue acts as an inadvertent secondary combustion chamber, as too-high flames lick out of the stove. This is dangerous. Not even heavy-duty stovepipe is built to hold a fire, and live flame in the flue can ignite built-up creosote. So add on the secondary chamber. A steel drum with stovepipe-sized tubes of the thickest gauge steel your machine shop can bend, nickel-welded onto openings front and back makes a good secondary chamber. In the end nearest the stove, drill a one-inch-diameter hole. Take a three-inch-diameter circle of metal, cut a notch an inch deep and wide in one side and drill a stove-bolt-sized hole in the center of the disc and another just to one side of the hole in the drum. Turn one of the disc's end points up for a handle, attach it, and you have a secondary draft control. Move the disc so the notch uncovers as much or little of the hole as you want. Depending on your fire, you may want a bigger opening, or two, one on each side. Attach the drum to the stove and you're in business. For stoves venting from the back, the secondary chamber will probably have the attachment tubing put on the ends. For a potbelly or other stove that vents from the top, you may want to put tubes on the barrel's sides, the draft control on the end facing the front of the stove.

For horizontally designed—or any—home-built drum stoves, you can attach a second drum over the first as in making a modern "dumb stove." I'd support the top drum with an angle-iron bracket at the rear, have the connecting tube or tubes toward the front, and the secondary draft in the front end of the top drum. Then keep the hot fire well back in the bottom drum,

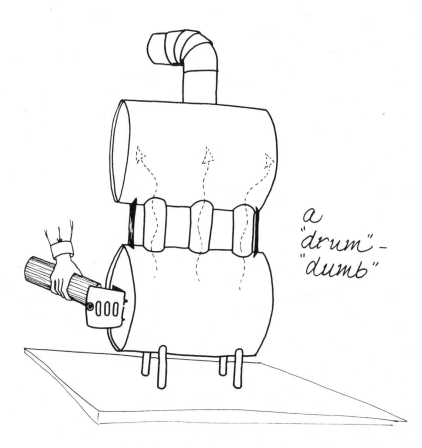

a "drum"- "dumb"

oven can be
built in
here

vent the top one from the back side, and you will be forcing the
flame into a C-shape as it travels through both barrels. This arrangement and a high fire can singe your whiskers from six feet
away.

**More Fire Safety Tips**

A hot fire is fine as long as the live coal bed doesn't get thick
enough to generate enough heat that it will burn through the
stove, warp plates, ignite creosote in the pipe or kindle any
nearby flammable material. If you need a hot fire, be sure to set
the stove up on several layers of brick or other noncombustible
base and keep flammable materials of any sort a yard or more
away from it in all dimensions. Also, if you are running any
length of stovepipe inside, secure it extra well and clean it
frequently. You can smell an overheated pipe or overfired stove;
it has that hot-iron sort of industrial smell that Louise and I first

noticed at our initial near-disaster back on the farm. To keep
your stove operating safely, keep the fire within bounds by closing damper and door to restrict air flow, and don't let the coal
bed build up by continually adding new wood. Let coals die
partway down once in a while.

When maple sugaring in late winter, cutting wood in fall and
during other cold-season activities out on the woodlot, we've
found that each time someone comes into the shack to warm up,
he fills up the stove. With a half-a-dozen people tossing in wood,
that coal bed can get up—perhaps when no one happens to be
around to watch. It's best to let a single individual run the fire if
possible, at least until everyone is accustomed to the stove. And
as we said in the beginning, never use cold water to calm an
overheated stove. Hot water isn't the best answer either. Close
off the air supply and flue and let the fire burn itself out. Save
your water to cool down any walls or other flammables that may
have been brought close to tinder point by a too-hot stove.

While we're stopping a stove fire safely, I should inject a
safety rule about starting one. Never use gasoline, charcoal
starter or whatever to get an inside fire going. They are okay for
the outdoor barbeque only. Never use them inside. And
particularly never add, say, a squirt of lighter fluid to an ornery,
inside smouldering fire that won't seem to get going. It can flash
back on you or hit the fire, vaporize instantly in the heat and
ignite with a really big bang. It's dangerous to have any flammable liquid in the same room with a wood fire, in my opinion. I
don't even like those Cape Cod fire lighters that are becoming
popular again. They have a chunk of porous rock on a handle
that sat in a pot of whale oil in the old days. Now it's kerosene or
charcoal lighter. The stone soaks up the stuff, you build the fire
over it and light the stone, which burns off the liquid slowly, giving the fire a slow steady start. You take the lighter out when the
fire is cold. Well, it does make things easy and the little brass
gadgets are an attractive ornament alongside stove or fireplace.
But if an errant spark happened to pop out as you were adding
wood and arch over into the pot or if a child, temporarily
transmogrified into a P-51 Mustang on a combat mission happened to zoom by and kick the pot into your Franklin . . .
Well, I don't like them! Besides, a proper, "for-heat" wood fire
only gets lit once a season. You just keep adding logs. Who

*for cleaning out ashes...*

*... a coal scuttle 'n shovel*

wants a lighter buried in the ash bank all winter? It would burn up, most likely. Or who wants to pick it out, sizzling hot when you are removing ashes?

### Ashes, Soot and Accessories

As for the ash, you will have some 50 to 60 pounds of it to deal with per cord of wood you burn. You'll want a coal scuttle and a long-handled flat shovel which you find in fireplace tool sets. With stoves having grates, you just shovel it out from under anytime it builds up to a level near the bottom of the grate. Most grated stoves are designed to have air coming up from the bottom, and if you clog them, the fire won't burn properly. Some model airtights, the circulators mainly, have ash drawers you just pull out. With a going fire, the ash will be hot. It is also light enough that it flies easily, so shovel and dump gingerly. With any stove in constant use, the dustpan and whisk broom are as important accessories as the tongs and poker—more important really.

In most grateless stoves, you really need an ash rake, a piece of sheet metal maybe four inches wide and two inches high attached in a **T** to a rod that's a handhold longer than your stove is deep. You may have to make your own the right size, especially for a big homemade drum stove. The foundries making cast-iron

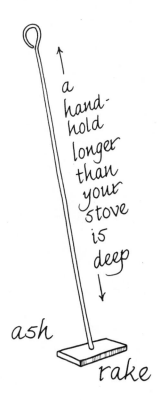

a hand-hold longer than your stove is deep

ash rake

kitchen ranges supply rakes that will suffice for all but the largest stoves. With a really big stove, you may be able to use a rake from an old coal furnace.

With tall fires in upright stoves, the pressure from constantly added wood and a little rake work will get the ash moved down. In stoves that burn wood lengthwise, you'll be pulling coals from the last-burned back end of the logs up to the front each morning and when you stoke up for the night. Just shovel the excess ash out into the scuttle.

If you use old-fashioned wooden matches, get an old-fashioned matchbox holder and nail it high on the wall out of reach of children, mice, sparks and heat from the fire. And while we're on fire-tending accessories, one item you don't need is one of those gadgets to "turn newspapers into fireplace logs." For one thing, most folks come equipped with a dandy set of paper rollers, one at the end of each arm. The crank or roller devices are not only unnecessary, they take longer and don't work as well as your own two hands. The rollers come with several

dozen lengths of twist-ties for holding rolled paper logs together. They sell extras for something approaching a dollar a
dozen. The ties are nothing but cut-up lengths of the same wire
and paper "twistems" we use to hold tomato plants to stakes in
the garden. If you must burn newspaper (rather than recycling it
into good things such as this book) you can buy a quarter-mile of
uncut twistem for the same price as you'd pay for a set or two of
"official" paper-log ties.

And then, newspapers don't really burn very well unless
they are so loosely rolled that they go up in seconds. You must
have a good bed of wood coals to burn well-rolled paper logs,
and once they are burned to charcoal, you must crush the
blackened, half-burned logs down into the coals to force them to
finish combustion. I'm told that if you roll logs good and tight,
but with a tubular opening through the center, then soak and
dry each one, the paper will compress into a fair approximation
of wood. We've never tried it and probably never will as long as
our town has a paper recycling center.

If you do burn paper, stick to newsprint or plain kraft paper—the brown stuff shopping bags are made of. Most magazines are printed on coated stock, which is covered with clay and other finishing agents that won't burn, and even if you poke them enough to get partial combustion, they'll fill your ash pit with chips and flakes of unburnt paper.

### Reclaiming Heat from the Stovepipe

The smoke and gases leaving a stove can reach 1,000 plus degrees F. and that heat should be salvaged if possible. The extra-long stovepipe just radiating heat into the room is a good measure, as mentioned earlier, though it must be kept clean and

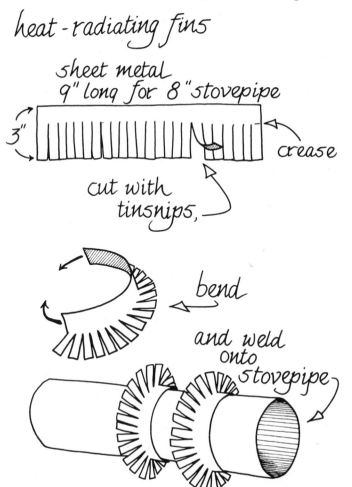

heat - radiating fins

sheet metal
9" long for 8" stovepipe

3"

crease

cut with
tinsnips,

bend

and weld
onto
stovepipe

well guyed. Several firms are making safer kinds of pipe heat reclaimers now, and more are sure to come on the market. The
simplest does nothing but increase the heat-radiating area of
exposed stovepipe by wrapping it in strips of aluminum crinkled
into a continuous M-shape. You can buy a set for a ridiculously
high price, or you can make your own from strips of aluminum
flashing obtainable at any hardware store. More effective but
none too attractive would be to weld on fins per the illustration.

There are two basic designs of mechanical heat reclaimers,
stack robbers or whatever you want to call them on the market
now (alluded to earlier). The most sophisticated uses a space-age
heat pipe that employs a fluid circulating inside a closed tube to
pull heat out of the flue and into the room. One unit I know
combines these pipes with a fan in a housing that fits right into
the stovepipe. The other design mentioned earlier runs tubes
through the flue gases and blows air through them. Electricity
used by the fans is negligible and all can increase the heat output
of your wood pile.

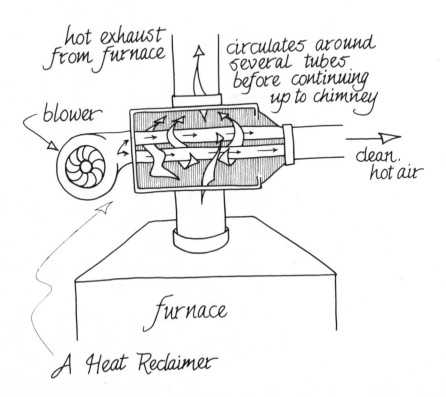

A Heat Reclaimer

By now competition has developed to the point that prices of the simpler types have gotten reasonable. Still cheaper is my favorite stovepipe heat exchanger, the stovepipe oven we put on the oil-drum stove. It even has a heat indicator in the door. Stick it in the stovepipe, put a damper above and below. Leave the door open and you've an excellent heat reclaimer. There is even a scraper inside the hollow donut of the outer and inner liners. (Regular reclaimers have one too; have to.) Before firing up each morning, you just give the oven scraper a couple of turns to dislodge any soot that gathered overnight. And with a bit of experimenting, it will turn out a great stew, pie, or loaf of bread.

# CHAPTER FOUR

# Fireplaces, Old and New

About the same time that Benjamin Franklin was improving the heating efficiency of fireplaces by turning them into stoves, another Ben, one Benjamin Thompson, was calculating how to improve the fireplaces themselves. A remarkable man—like Franklin an inventor, diplomat and man of letters—Thompson unfortunately failed to side with Franklin and friends in the Revolutionary War and in 1776 was sent packing to England and out of American history. He did well for himself, though; earned the title of Count and took the name of his wife's hometown, Rumford—now Concord—New Hampshire. The improved firebox and flue design Count Rumford perfected is known as the Rumford fireplace, and his basic dimensions have never been improved for a good-heating brick-and-mortar fireplace. The illustration "Comparison of Fireplace Designs" shows and explains the difference in vertical proportions between a Rumford and a modern recreational fireplace. The great high mass of masonry in the Rumford absorbs and the angled back reflects most of the heat from the fire. Stand in front of one and you feel the gentle radiation six or more feet away. The combustion gases have lost their heat by the time they reach the lintel, and are just warm enough to keep the flue drawing. The modern fireplace, on the other hand, is designed to waste fuel as efficiently as possible. The shallow, low firebox effectively puts most of the heat up the chimney. There's even an ash dump at the back so you can keep the firebox "clean" by shoving the hottest part of your fire down into the cellar.

## Fireplace Basics

We touched on principles of fireplace operation earlier. Recall the American Indian tipi, where air for combustion and a

Here is an antique Rumford-design fireplace in the home of Howard Hastings of Barre, Massachusetts. There is no damper, and until one is installed, Hastings stuffs fiberglass insulation into the throat when the fireplace is not in use. Several of the fireplaces in this old home still do have the original "dampers," wooden screens that fit snugly into the firebox framing to keep in warmth when the fire was out.

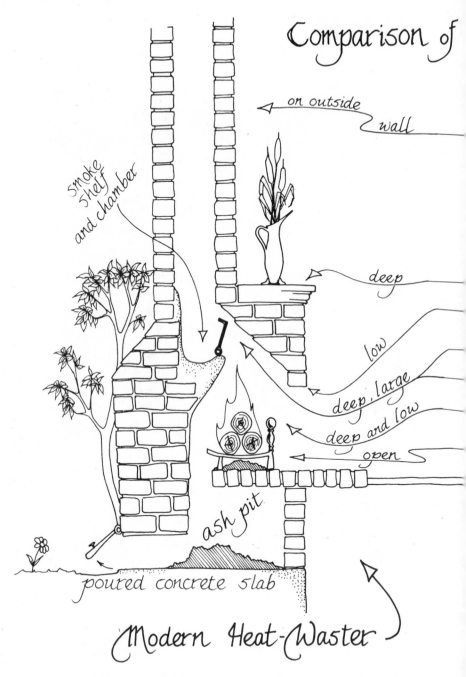

Comparison of

on outside wall

smoke shelf and chamber

deep

low

deep, large

deep and low

open

ash pit

poured concrete slab

Modern Heat-Waster

You'll note that both these fireplaces have smoke shelves and smoke chambers, so neither will smoke badly if proportions are correct. The big difference is in the size and shape of the firebox. In the modern design, all the heat is sucked right out of the low little box and up the flue. In the Rumford, heat is absorbed by the masonry all the way from the solid ceramic base to the bricks in the breast up to the lintel. There is at least twice the heat-absorptive mass and

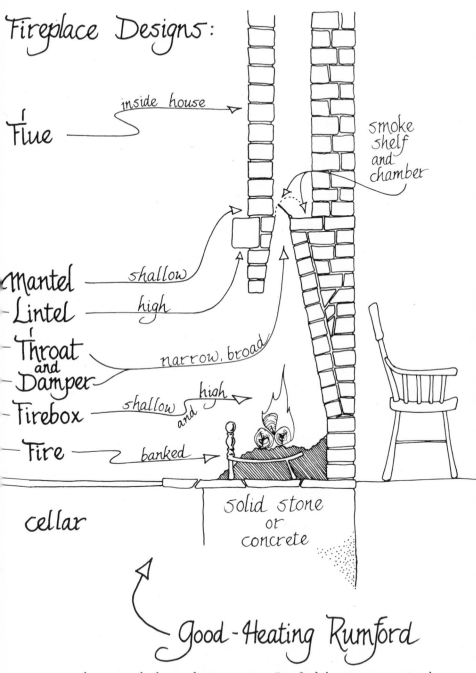

**Fireplace Designs:**

Flue — inside house

smoke shelf and chamber

Mantel — shallow

Lintel — high

Throat and Damper — narrow, broad

Firebox — shallow and high

Fire — banked

cellar

solid stone or concrete

**Good-Heating Rumford**

more than twice the heat-radiating area in a Rumford than in a conventional fireplace. Further, with the great height of warmed masonry at the firebox back, combustion gasses will be drawn into the flue from well out in front of the lintel. And many users have their fire fronts a foot or so out into the living space. If you are going to build a new brick-and-mortar fireplace, make it a Rumford.

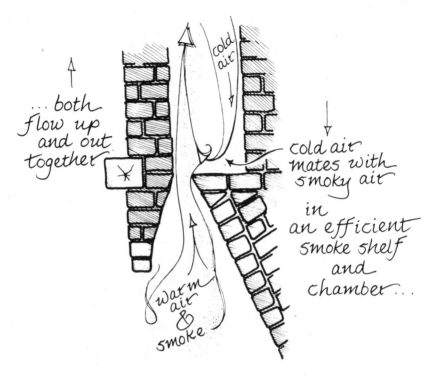

... both
flow up
and out
together

cold
air

cold air
mates with
smoky air

in
an efficient
smoke shelf
and
chamber...

warm
air
&
smoke

healthy atmosphere came in through small openings at the bottom of the tent. Draft openings in what was a giant fireplace/flue, in effect. And the top flap acted as a damper and smoke shelf, being regulated so that air flowed in at the top and circulated down to be warmed, then took the smoke out as it rose. This is how an efficient fireplace operates, or how it should. Ideally, only enough air is drawn in to remove combustion gases from the firebox. David Howard, an architect and builder from Alstead, New Hampshire, who puts Rumford fireplaces in his post-and-beam houses, has a simple test of a fireplace's drawing power. If the smoke from his pipe, released at the lower edge of the mantle, is drawn gently, slowly but completely into the flue, the draft is right.

If the firebox lacks a smoke chamber and empties right into the flue, as is the case in many older fireplaces and, unfortunately in some new prefabricated designs, you may have a smoky fireplace. That cold air that pours down the flue will roll out into the room, carrying smoke with it. So in Rumford and other efficient designs, the smoke chamber has a pronounced shelf at the bottom just above the firebox. Outside air flows down, hits the warm shelf and mates with the smoky air entering from the firebox, then all flows happily up and out.

For proper functioning, you must maintain strict ratios between size of fireplace opening, size and height of flue and so on. The figures are given in the following pages; deviate from them too much and you are asking for trouble. To explain what problems you might bring on yourself, let's examine the reasons an existing fireplace might not perform properly and suggest solutions. Use any ideas that fit problems you may have with your own.

patching
up
the
firebox

**Inspecting the Facility**

Before lighting a first fire, check the fireplace for bad joints, cracked bricks and all, just as we looked the flue over earlier. The inside of the firebox may be of brick or soapstone or other type of rock slabs if it is an older one. Circa 1920 fireboxes are often lined with large fireclay blocks and in newer homes you'll usually find a steel liner. Check joints in all ceramic boxes and patch any small holes with furnace cement. Remove any loose bricks and replace with fireclay mortar.

If you've a metal liner, pound around looking for holes. You may be able to patch a small hole with furnace cement or put in a bolt or screw to seal it up. If there are big holes or if the damper mechanism is rusted through, call a mason. The problem with the metal liners is that they do deteriorate in time, particularly the less-well-made ones, and repair is difficult. You may have to have a new firebox put in. Make it a brick-and-mortar Rumford—coming up a bit further on.

With all fireplaces, get your hand as far up into the flue or smoke chamber as possible. Remove any fallen bricks, birds nests or whatever. Here is one advantage of modern design; the big opening gives better access to the smoke shelf than the narrow Rumford throat.

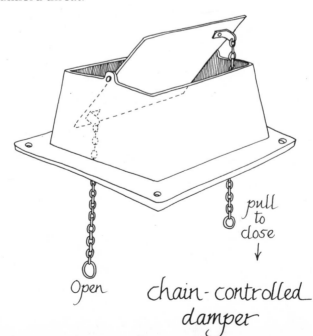

pull
to
close
↓

Open

chain-controlled
damper

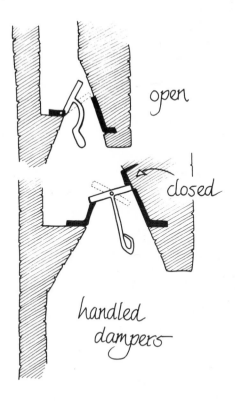

open

closed

handled
dampers

Hopefully, there will be a fireplace damper in the throat, a metal door that closes the flue when the fireplace is not in use. Be sure it opens and closes tightly. It should be hinged at the back and the lever, handle or other control should open it so the metal plate rests fully vertical. If there is no damper and you live in a chilly climate, call the mason and have one installed. Until it is in, stuff rags into the throat when the fire is cold, else all your warmth will go up the flue.

## Lighting Up

You should have a set of andirons or log rests; if not, get a pair that extends the full depth of the firebox. If you have a fire grate or box arrangement it was most likely designed to burn coal and will not hold a properly banked wood fire. Let's build one. Open the damper all the way and wait a few minutes, particularly if the weather is cool and you have a typical modern chimney, built on the outside of the house. Cold chimneys don't draw, remember. You may have to warm it with a small hot

good-sized, dry logs

criss-crossed kindling

loosely crumpled newspapers

Laying a proper Fire

wood fire or a few sheets of paper lit and held high in the fire-place.

And remember to let the wood warm too. If you try to start a fire with frozen logs just brought in from outside you'll have a long wait. The kindling in particular should be room temperature. Crumple up several sheets of newspaper—a loose crumple—and put them on the firebox floor. Take a good-sized, preferably unsplit log, lay it on the andirons, and push it all the way to the back of the firebox. Put another big one, preferably split out in front and fill the space between them with kindling sticks laid so plenty of air can get through. Then on top of the kindling lay another split log. Arrange the three large logs so they are almost but not quite touching along most of their lengths, in Louise's magic wood spacing arrangement. Just as it did in the stove fires, a narrow opening between logs increases velocity of the draft flowing up from underneath, in effect "blowing" on the fire to increase oxygen supply. Not an effect you want in a good heating fire, but necessary to get the blaze going.

Now, light the paper. If you've a good draft and the wood is properly seasoned, split the right size and warm, you should never have to use more than two sheets of newsprint to start a fire—if you've got the knack, that is. I can still go through half the Sunday paper in front of a fireplace at times.

### Keeping the Fire Burning

Once the fire is going, your object is to create a heavy bed of ashes; indeed the entire floor of the firebox, at least up to the top of the andirons, should be solid with warm ashes. The heart of

red-hot coals...

... heart of a well-banked fire.

the fire is a body of red-hot coals in the center of the bank that slowly settles down as it goes to ash while radiating heat with maximum efficiency. Indeed, once you've the fire established, you should seldom see a flame more than a few inches high. Logs go on one or two at a time and are continually pushed back into the ash bank. Remove excess ashes as they are pushed out onto the cool front of the bank and to the sides. This way you can light up once in the fall and keep the same fire till spring. Of course, it will take some time to build up the first ash bank and if you let it get cold, you'll have to start over each time with a new fire. However, if you're seriously heating with wood, the well-banked, smokeless and nearly flameless permanent fire is the kind to have.

Incidentally, there exists considerable difference of opinion around the country on just what a well-banked fire is—indeed what the term "banked" means. According to the dictionary it means to cover the fire with fresh fuel and reduce your draft to put the fire or coal bed in an inactive state. Well, hereabouts, the bank of ash-covered coals is part of a well-banked fire. You build up that ash bank, use the ashes to build a bank (the word also means a mound, as of earth or ash) over and around the coals, then shore up the ash bank with another load of logs. If you closed the damper down and told the fire to remain inactive, as the dictionary says, your house would be smoky pretty quick. Fires can't read—or hear. But they sure can heat if you build them right, banked three ways, I guess. (That dictionary definition must be a semimodern reading of an old term. You could do as it says in a coal furnace, a foundry or in the engine of the old Wabash Cannonball. But not in a fireplace.) The banked fire is vastly more efficient than any ornamental blaze (though there's no harm in tossing on a few small logs for a cheerful flame in the evening). But, in the ash bank your logs will turn to coals from the bottom up, burning completely but slowly so that the maximum amount of the heat stays in your room and none is lost in those cheerful, but wasteful high flames.

## The Smoky Fireplace

Now, what if your fireplace smokes?

The most common cause of smoking in newer houses is a lack of air for the updraft. Turn off the kitchen or bathroom

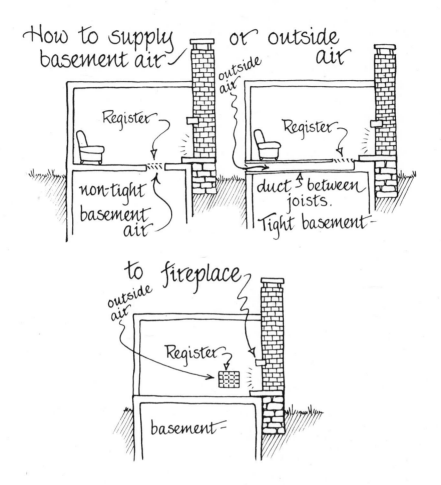

How to supply basement air or outside air

outside air

Register

non-tight basement air

Register

duct between joists.

Tight basement

to fireplace

outside air

Register

basement

exhaust fans and open a window in the next room a crack. If the resulting draft through the room and along the floor is a nuisance, you might want to drill holes in the wall on both sides of the fireplace. There are small, closable registers for such use. By the way, such air inlets can improve the heating efficiency of a conventional fireplace in almost any house. It isn't widely known, but a roaring fire with damper full open in the usual recreational fireplace can draw out more centrally heated house warmth in its draft air than it contributes through radiation. (Not so in a Rumford.) You can reduce the loss (and reduce smoking as well) in any fireplace design by putting air inlets into walls beside the fireplace, drilling holes in floor or hearth in front of the fire or building a steel conduit up the back of the firebox to fit into the opening of the ash-pit; then outside or cellar air will pour up the flue, reducing room draft.

With a good air supply, and smoke coming in puffs, not continuously, you've a chimney problem, most likely an inability to handle strong wind gusts. Either install a wind barrier (see page 46), or raise the flue (see page 221). A minor, but naggingly constant bit of smoke in the air can often be eliminated with a smoke curtain or hood that effectively drops the lintel. An old, but dangerous practice is to hang a thick cloth from the lintel. A metal strip, hood shaped or simply flat, can be

a
lintel-lowering hood...

... to remedy
a
smoky fireplace.

mortared in place—and won't burn. Perhaps after experimenting with a strictly temporary fire or smoke curtain, you might
want to put in another lintel support and drop the masonry
several inches.

If you are still choking on smoke put out the fire and
measure all dimensions of the fireplace, flue and smoke chamber
and compare them to the figures in the chart. Most commonly
you'll find the throat too small or too large and situated too far
back in the firebox. It should be at the front and its area (length
times width) should be at least equal to the flue area. With a too-
big throat and an adjustable damper, you may be able to choke
the throat down enough to improve the draft. If the firebox

A lowered lintel...

...and added smoke shelf with damper

opening is too large for the throat/flue relationship, you can modify the dimenions with firebrick at the back, sides or on the bottom. Before actually mortaring in any changes, I'd experiment by just laying up regular building brick and trying out the new dimensions with a small, but smoky fire. Often simply raising the hearth can help a lot. That is what we did back on the Pennsylvania farm, where the fireplace had been built without any smoke chamber at all. Raised a foot, and with a foot-wide black iron curtain to drop the lintel, the smokiness just about disappeared. To have attempted to reconstruct the fireplace and put in a proper smoke chamber was just too much of an undertaking for our limited know-how at the time, and I'm still not sure of how to go about such things as supporting the flue top while reworking the bottom. Had we done it, a professional mason would have been called in, and I'd advise you to do likewise if relatively simple measures fail to cure your smoky fireplace.

One other corrective measure we have learned about since our smoky fireplace experience is to lower the lintel if needed and mortar in a sheet-steel angle at the back of the fireplace, in effect making your own smoke chamber. This would probably not be possible in a small fireplace and I don't know but that the aesthetics would prevail over a bit of smoke in the air with an antique walk-in. But the drawing shows what I'm getting at and it may work for you.

A common problem in old-but-not-antique flues are cracks or open joints in the wythe, the wall separating your fireplace flue from a central-heating flue. Air passing between flues will interfere with draft in one or both and the traveller and cement treatment described earlier will help. Other flue problems can include too-steep angles, bumpy flue linings and improper height. There's not much you can do about any but the latter problem except rebuild.

Recall that the chimney should rise at least two feet higher than the roof peak or any tall, nearby objects. Add on another foot of height if your roof is flat or nearly so. Also, measure the area of the fireplace opening and height of the flue from the firebox bottom. The opening should be no more than 12 times the area of the flue's inside dimension if the flue is over 22 feet high. If the flue is less than 22 feet high, it should be at least

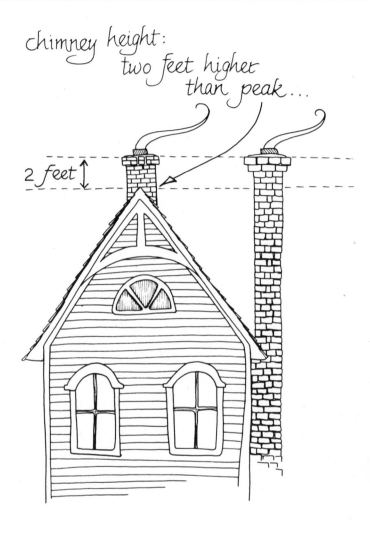

chimney height:
two feet higher
than peak...

2 feet

one-tenth the area of the fireplace opening. If the chimney is too low, build it up with bricks or whatever or tack on a ceramic smoke pot or steel rotating ventilator. A sheet metal tube just rammed down into the flue can help if you don't care much about appearances. For details, check back through the chapter on old flues.

*... three feet high on a flat roof.*

← → 3 feet

## Improving the Efficiency of Existing Fireplaces

Now that you've made sure your fireplace doesn't smoke overmuch, let's see what can be done to make it a better heater. As it is, I'd bet you are losing 90 percent of your heat up the flue.

One thing is to learn to use the damper. Most commonly, dampers come with a notched handle or other device for regulating the size of the throat opening. Get that well-banked fire going as far out in front of the firebox as you can and see how far down you can close the damper. If your damper only has two settings, open or closed, see if there isn't something you can do to give a series of intermediate settings by tightening bolts or filing notches in the handle. Experiment by stuffing asbestos board or brick into the opening to see if restricting the throat size is practical.

There are a number of patented grates and other devices on the market now for getting more heat out of the fire. One variation on the theme is made by the Dahlquists of 31 Morgan Park, Clinton, CT 06413. Their Radiant Grate raises the fire's skirts to expose coals to the room air, and they sell reflector ovens, a stove and a grill to cook by radiant heat. All their meals are cooked before the fire. And from the photos I've seen, their fireplace is a conventional fuel-waster—without the cooking feature provided by a grate, that is.

You've probably seen ads for tubular grates, a series of C-shaped tubes welded on legs to face out into the room. The idea is that cool air is drawn in at the bottom, rises as it is heated by the fire, then flows out to warm the room. Some come equipped with blowers to increase the air flow. If you don't mind a blast of hot air in your face each time you throw a log on the fire, they work. But, beware if you buy one. Many are coming out made of nothing but auto exhaust pipe painted black to resemble cast iron. I don't know of any that come in real cast iron, but shop around for the heaviest gauge pipe you can find. The thin stuff won't last through two heating seasons. And I'd think that the ones with a blower would last longest as the greater air circulation would keep the pipe cooler than natural air flow.

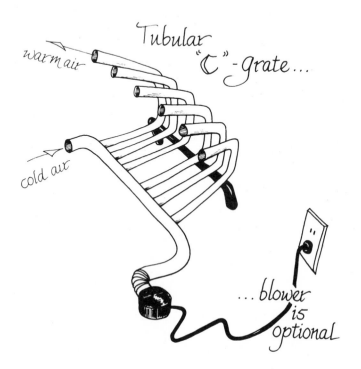

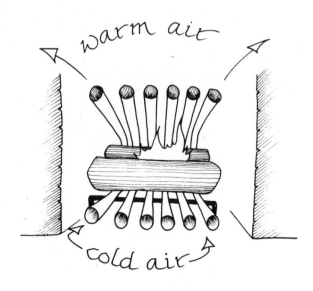

*... another model*

As wood heating increases in popularity, more and more new ideas will be coming onto the market. I've seen ads for other devices that run air through the top or bottom of the firebox—big horizontal metal tubes formed in a C-shape, or serpentine arrangements with a fan at one end and a lot of hot air at the other. These at least improve on the grates made in a series of vertical C's, which effectively blow the hot air right up to the ceiling.

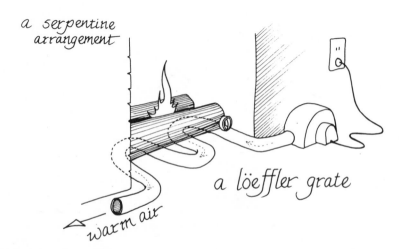

Any of the patent grates will run $50, more or less, and the electricity the blower-equipped models use is negligible compared with the Btu's of heat they keep from being wasted up the flue. I'd say that any well-made one is a good long-term investment if you are stuck with a conventional fireplace. Still, none of them will make a heat-waster into a Rumford.

Most fireplace accessory stores sell fire screens made of wire mesh and others of glass. I don't like the mesh between me and the heat, but I wouldn't leave the room for long with a fire going without one. (And neither of us would leave the house with a live sparking fire going under any circumstances. A well-screened ash and coal bank with virtually no new wood on top— a solid ash covering, but enough coals beneath to hold heat for hours—is safer.) But those glass doors are calculated to turn a normal wood-wasting fireplace into a super squanderer. Draft is brought into the fire through slots at top and/or bottom, so room air loss is minimized. Glass *does* radiate heat, but not nearly as well as plain air. As with the airtight stoves, I'd recommend regulating air flow at the flue, not at the draft opening. Plus, even good "tempered" glass will crack when large sheets are exposed to uneven heat. Any glass exposed to an honest heating fire should be cut in small pieces—either framed in metal as in the old illuminated stoves or several sheets butted together in a metal frame as in the circa 1920s kitchen ranges' see-through doors.

## Building a Fireplace

Fireplace construction is about the ultimate skill in the mason's trade, and trying to explain it in detail is beyond the scope of this book. It's also beyond the scope of this writer's skills and first-hand masonry experience. Stone walls and flue maintenance and such simple stuff as mud-and-stick or precast cement-block flues I've done. But take a hard look at a brick and mortar fireplace. The inner surface of the firebox, smoke chamber and flue should be perfectly flat and smooth so smoke will flow freely. And look at the angles of the firebox's back walls. The back slopes up, out and toward you. The two sides do the same thing but at a biased angle. It takes a great deal of time and the advice of an experienced mason. Bricks have to be beveled at angles in the box and woven where walls meet, and

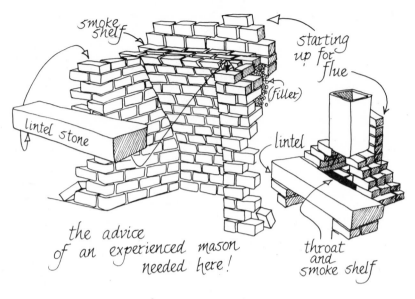

smoke shelf

starting up for flue

(filler)

lintel stone

lintel

the advice of an experienced mason needed here!

throat and smoke shelf

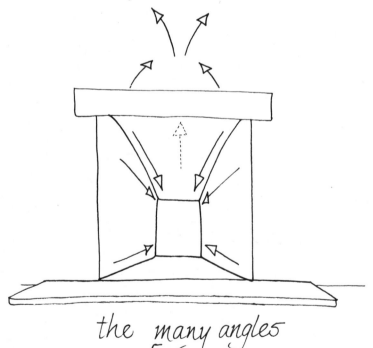

the many angles of a fireplace

everything has to end up plumb, square and smooth. An amateur *can* do it, but not by simply reading a book.

What attempts I have seen in books to describe fireplace construction are nothing but paraphrases of a government publication—and it tells you on page one to get an experienced

## DIMENSIONS FOR CONVENTIONAL MODERN FIREPLACES AND FLUE SIZES

(Letters at heads of columns refer to drawing below)

| Size of fireplace opening | | | Minimum width of back wall | Height of vertical back wall | Height of inclined back wall | Size of flue lining required | |
|---|---|---|---|---|---|---|---|
| Width | Height | Depth | | | | Standard rectangular (outside dimensions) | Standard round (inside diameter) |
| w | h | d | c | a | b | | |
| Inches | Inches | Inches | Inches | Inches | Inches | Inches | Inches |
| 24 | 24 | 16-18 | 14 | 14 | 16 | 8 x 13 | 10 |
| 28 | 24 | 16-18 | 14 | 14 | 16 | 8 x 13 | 10 |
| 30 | 28-30 | 16-18 | 16 | 14 | 18 | 8 x 13 | 10 |
| 36 | 28-30 | 16-18 | 22 | 14 | 18 | 8 x 13 | 12 |
| 42 | 28-32 | 16-18 | 28 | 14 | 18 | 13 x 13 | 12 |
| 48 | 32 | 18-20 | 32 | 14 | 24 | 13 x 13 | 15 |
| 54 | 36 | 18-20 | 36 | 14 | 28 | 13 x 18 | 15 |
| 60 | 36 | 18-20 | 44 | 14 | 28 | 13 x 18 | 15 |
| 54 | 40 | 20-22 | 36 | 17 | 29 | 13 x 18 | 15 |
| 60 | 40 | 20-22 | 42 | 17 | 30 | 18 x 18 | 18 |
| 66 | 40 | 20-22 | 44 | 17 | 30 | 18 x 18 | 18 |
| 72 | 40 | 22-28 | 51 | 17 | 30 | 18 x 18 | 18 |

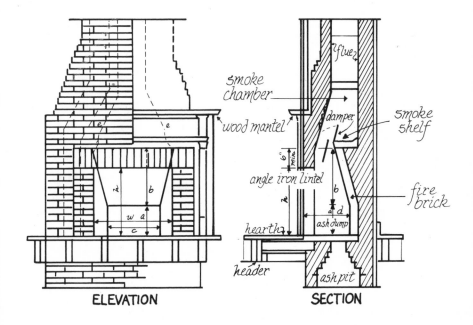

ELEVATION          SECTION

This is an example of colonial restoration and fireplace construction at its very best. Mason Jim Dowd is constructing a late 1700s kitchen fireplace, complete with beehive oven.

mason. The booklet is "Fireplaces and Chimneys" (Farmers Bulletin No. 1889), issued by the U.S. Department of Agriculture (from the Superintendent of Documents, Washington, DC 20402). You will also find a selection of books on the trowel trades in most public libraries.

However, no books I've seen adequately describe the dimensions and structure of a good-heating Rumford-style fireplace. You can get the Count's journals which ramble on about it in eighteenth century English in a few public libraries (Harvard House is the publisher). Or you can send $2.50 to Yankee, Inc., Dublin, NH, for *The Forgotten Art of Building a Good Fireplace*, by Vrest Orton. The book contains a lot of philosophizing that some folks might think more appropriate to Count Rumford's time than our own, though Rumford's basic fireplace dimension principles are there too.

Rumford insisted on characteristics that modern builders have found unnecessary, such as requiring that a plumb line dropped from the center of the throat land squarely at the center of the firebox. Presumably that perpendicular should go straight up the flue, so smoke would also go straight up without hin-

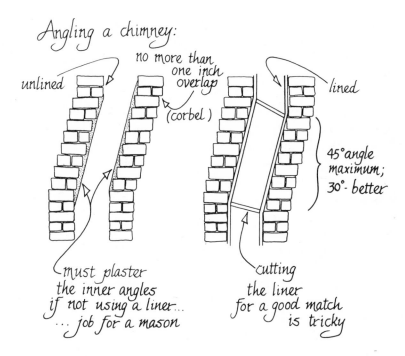

*Angling a chimney:*

unlined

no more than
one inch
overlap

(corbel)

lined

45° angle
maximum;
30°- better

must plaster
the inner angles
if not using a liner...
... job for a mason

cutting
the liner
for a good match
is tricky

drance. I wonder if that rule was Rumford's way of getting people to build in a minimum of angles in a flue; by his rule you could look up the throat and see the sky, which is the case in some of the best-drawing antique fireplaces I've seen. However, modern masons know you can put in as much as a 45 degree angle if absolutely necessary, though chimney height may have to be increased to compensate. Better is to keep angles down to 30 degrees or less.

### Planning for a Good-Heating Fireplace

Though I'll not attempt to tell you or your mason how to build a fireplace, I will offer the basic principles that Louise and I considered when trying to decide between a fireplace and a stove to heat our present house. Perhaps you'll find them useful to your own planning for a new house (build the house around the fireplace) or installing one in an existing dwelling.

First, the fireplace and flue are the heaviest part of any home and if you go with masonry from cellar to roof peak, the structure needs its own foundation. An eight-inch-thick, poured concrete slab is usual in modern construction (where tract houses are built to last only 15 years). If yours is an older house, like ours, built to last, or if you are building one to last, you would want a solid pier of cemented rock, concrete block or poured concrete. Make it a foot around larger than the base of the fireplace and any other flues. Have it extend well below frost line. Putting a new fireplace into an existing house means sawing a big chunk out of your floor beams and surfaces.

Next, if you are serious about heating with the fireplace, you will want the entire brickwork within the house spaced so heat will radiate out from all sides. As I've said before, putting the fireplace and flue on an outside wall is easier to build, especially in an existing structure and we put one in ourselves. But for heating, it is wasteful, and if you can keep chimneys inside, you should do so.

I mentioned the 30 degree optimum bend that you could put in if needed to get the flue around plumbing or structural items that you can't or don't want to move. Another rule is that the masonry must never be closer than two inches from any flammable material, and that includes all beams and rafters. Then there is the problem of getting the chimney works through the upper floors and the roof, each level requiring a major rebuilding job. As it turned out for us, using the new flue location that made best heating sense in our house (the center alongside the existing flue) would have forced us to cut through all main supporting timbers from cellar to roof. Too much.

What we did consider, briefly, was a compromise all around: a good-heating masonry structure from floor to ceiling in the living room, and prefabricated stovepipe on up through the upper stories and the roof. Recalling the old plank-supported fireplace on our first farm and after talking with a house mover, we planned to support the mass as follows: sister beams along existing joists, six-inch-thick oak planks bedded in a leveling layer of mortar under these, held up by ten-ton house jacks (six, probably) set atop a concrete-block pier. Or at least that was the only way I could figure how to do it myself without having cement mixers chewing up the lawn.

The only problem we didn't get solved to our satisfaction, and a question that surely would have taken the advice of a builder and probably the fire marshall, was how to prevent the flooring from catching fire. Perhaps a high hearth set on perforated brick (building brick with two or three holes in it) or porous cinder block would have done it. I considered laying perforated brick with holes aligned all the way through and fixing a blower to one side, keeping the brick cool and dispersing heat (perhaps through piping to the upper floors). Though taking out floor beams would have been too much of a job, we could have easily torn up the flooring and replaced it with a non-combustible, nonconductive material, most likely fine quartz sand.

Let me reiterate: We have not yet done this and don't know if it would work or satisfy our building codes. I make no recommendation for it. However, everyone these days is coming out with ideas on how to get the most out of wood heat, so I thought I'd offer one of my own. If you want to try it, I should pass on the evaluation of master mason Jim Dowd, a Rumford builder from Hardwick, Massachusetts: "Too expensive, impractical and won't last." A man of few words, Jim is rebuilding an old colonial fireplace that rests on rotted timbers, and he oughta know.

### Designing the Fireplace

Back to Mr. Thompson, or Count Rumford, and his fireplace. Assuming you have your foundation set, mason with trowel in hand, what's next? First, decide the general dimensions of the finished fireplace, hearth and trim. Interior design is another topic that is beyond me, but don't think you have to stick to Early American decor just because the fireplace is based on old principles. It will look fine in any setting. There are several good books out on fireplace design. All contain more or less the same technical information you'll find in the USDA manual. None I've seen will help you heat with wood, but they do have pictures of all styles of rooms containing handsome fireplaces cheerily wasting wood.

We've combined Count Rumford's principles with the practical know-how of several men who are building modern versions of his fireplaces to compare dimensions of several Rumford-style fireboxes with a modern fuel-waster. The differences are not great, but they are crucial: the firebox shallower, higher and angled forward and lower down, the throat more narrow and extending all across the front of the firebox. Your mason will have his own ideas about size and shape

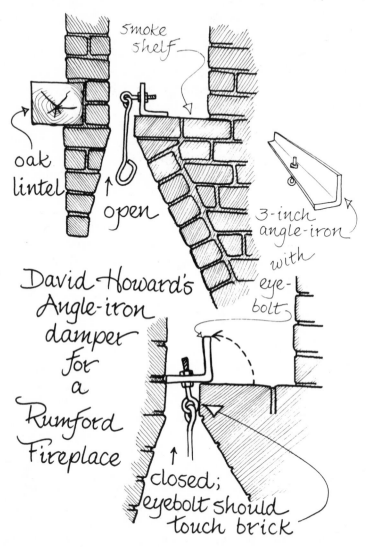

smoke shelf

oak lintel

↑ open

3-inch angle-iron with eye-bolt

David Howard's Angle-iron damper for a Rumford Fireplace

↑ closed; eyebolt should touch brick

of the smoke shelf, what bricks to use where and so on. Just make sure he sticks to a basic Rumford firebox. Oh yes, you'll find no damper made for a Rumford. David Howard (whose 40-by-40-inch design is included in the chart) uses nothing but a length of three-inch angle iron with a rod attached. Just flip it up or down for open or closed. And you can unhook your operating rod, turn the iron over and slip it down and out to get at the smoke shelf. We've included a sketch of a more conventional damper you can have made at any job shop. Good building.

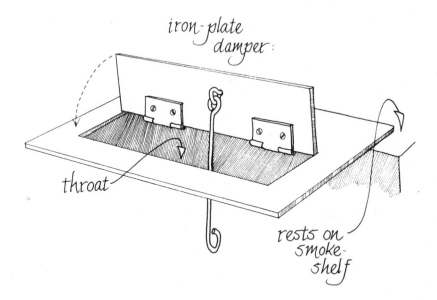

*iron-plate*
*damper:*

*throat*

*rests on*
*smoke-*
*shelf*

---

### To Build a Rumford Fireplace
#### Basic differences from conventional modern fireplaces

1. The throat is as wide as the firebox, but only from three to four inches in depth. (Some say to four-and-a-half inches.)

2. The width of the front opening is two to three times the depth of the firebox (usually three and always equal to the height).

3. The height of the front opening is two to three times the depth of the firebox (usually three and always equal to the width).

4. Height of fireback, where it begins to slope forward is usually 12 inches, less in small boxes.

5. Width of fireback is equal to the depth of the firebox.

6. Smoke shelf is as wide as the fireplace opening and 12 to 16 inches deep. (Some experts say 6 to 12 inches deep.)

7. Smallest permissible flue size is the larger of: one-tenth or twelfth the area of the fireplace opening depending on flue height or the area of the throat.

## COMPARATIVE DIMENSIONS OF
## FIVE FIREPLACES

(in inches)

| Type of Fireplace | Width | Firebox Height | Depth | Fire back FW. | FH. | L. | TW. | Throat TD. | Flue |
|---|---|---|---|---|---|---|---|---|---|
| **Rumfords** | | | | | | | | | |
| | | | 8 to | 8 to | 8 to | | | | |
| Small Upstairs | 24 | 24 | 12's | 12's | 12's | 8 | 24 | 3 | 8 x 9 |
| Yard-wide | 36 | 36 | 12 | 12 | 15 | 15 | 36 | 3 | 10 x 13 |
| Cordwood | 50 | 50 | 17 | 17 | 20 | 12 | 50 | 4 | 13 x 16 |
| **Howard** | | | | | | | | | |
| Yard-wide | 40 | 40 | 18 | 17 | 17 | 12 | 40 | 3 | 12 x 16 |
| **Modern** | | | | | | | | | |
| Conventional | | | | | | | | | |
| Yard-wide | 36 | 29 | 17 | 22 | 14 | 12 | * | * | 8½ x 13 |

*The throat dimensions of conventional fireplaces vary, but invariably the throats are too big or too small.

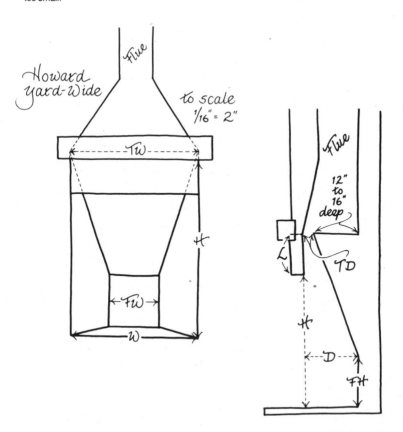

Howard
yard-Wide

Flue

to scale
1/16" = 2"

TW

H

FW

W

Flue

12"
to
16"
deep

L

TD

H

D

FH

Comparative - Dimensions Key

Here is a rear view of a Rumford-style firebox going up. Construction here is something of a compromise between modern and old-time method. The design is good old Rumford, but the material is modern 9-inch by 4½-inch by 2½-inch yellow firebrick, as used in most fireplaces being built these days. Note how even these large chunks of ceramic must be cut at often complex angles before being laid up with a thin joint of fireclay. It's no job for an amateur.

The firebox is almost complete now, the forward-leaning back face being supported by a simple wooden form. After the basic shape was laid, the entire firebox was surrounded by solid ceramic—here, concrete block and cement. This makes a great monolithic ceramic structure that will act as a perfect heat sink as well as a permanent base for the smokeshelf and flue.

And here is the completed firebox with the damper in place. An angle iron will be cemented in place about three bricks down from the top to act as lintel. Decorative brick will be laid up from the lintel and around the sides, the flue will go up and ol' Bob Furgeson will have himself a fireplace to ward off the hairiest winters the province of Ontario can throw at him.

# Prefabricated Fireplaces

As mentioned, a brick-and-mortar fireplace just wasn't practical for our place, so we halfheartedly looked over the prefab fireplaces. These are steel and come in two styles. One is freestanding, most typically shaped like a big tin funnel over a base shaped like a giant tuna fish can. The can is so well insulated you can set it in the center of your room as is, hang the funnel from the rafters, run an insulated flue up through the roof and have a fire that evening. Some of them come in bright colors that match modern decors, and I hear they are big in southern California. There they dig a firepit in the stone floor, surround it with a tier of seats, hang the funnel from a really big and powerful flue and toast their weenies. Of course, the things don't heat for a hoot, but who needs heat in southern California.

The other type of prefab, and the one we looked at, comes as a conventionally designed fireplace, complete from base to top of the smoke chamber. They are steel, multi-walled for perfect insulation and can be placed right on a wood floor, boxed in with two-by-fours resting right against the outside and covered with wallboard. Then you paste on plastic brick, connect your prefab flue and sit back to look like something out of the Sears catalog. If I'm being overly snide about these things, I apologize. They are cheap, quick to go up, and don't at all have to be faced with plastic, but can be bricked in or faced with stone or lovely colonial woodwork. They will look just like the real thing. But that's the problem. The older designs are all for looks. Especially the superinsulated jobs. They are so well insulated they are the ultimate in heat-wasters. They suck all the wood heat, plus your room heat, up the flue without even providing the consolation of a warmth-radiating brick firebox once the fire is banked and damper closed a bit for the night.

### Air-Circulators

The only prefabricated fireplaces that have any right to claim a genuine heating capability are the air-circulators. Again, two kinds. One is a standard steel firebox. You build a box, brick usually, around it. On each side, perhaps top and bottom if it is built into a flat wall, you install ventilators. Air flows in at the bottom, is heated by the firebox walls and comes out at the ceil-

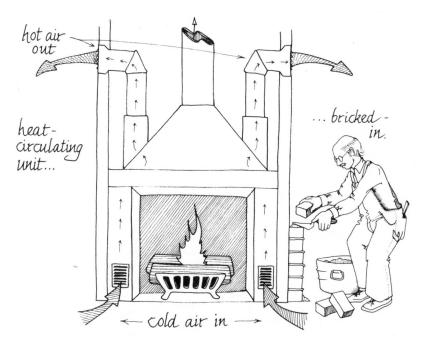

hot air out

heat-circulating unit...

... bricked-in.

← cold air in →

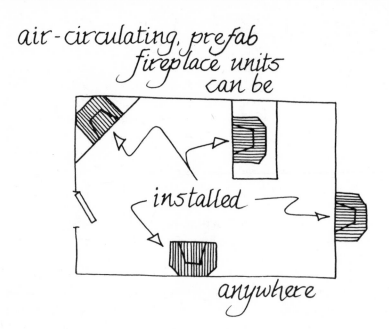

*air-circulating, prefab fireplace units can be installed anywhere*

ing. The other version is similar to the superinsulated standard variety, but it too has vents built in; air circulates around the multiple-layer firebox, comes out on top and is piped up to the ceiling. Some are designed like certain of the Franklin stoves to be bricked into an existing firebox.

Well, here again we have a blast of hot, dry air delivered in your face or at the ceiling where least needed. Worse, the fireboxes are standard wide-damper, deep firebox fuel-wasters. But worst of all, they can't hold an efficiently banked fire (and any brick-and-mortar fireplace can do that). All that air circulating around the thin steel sucks the firebox of its warmth. True, the circulated air warms your ceiling and perhaps the upstairs bedroom floor before it circulates around to be sucked out the flue, but only if you keep a great, roaring fire. Let it go to banked coals and shut down the damper and you'll soon have a dead fire. There will be a few sparks deep in the ash bank come morning, but the room will be mighty cold.

You see, the coals in an ash bank feed on themselves. They need to remain in intense heat to generate efficient heat, and that means slow heat produced with a minimum of oxygen. Just enough fresh air to keep their cheeks red. The ash bank acts as an air barrier and in a masonry fireplace, the brick is a heat sink,

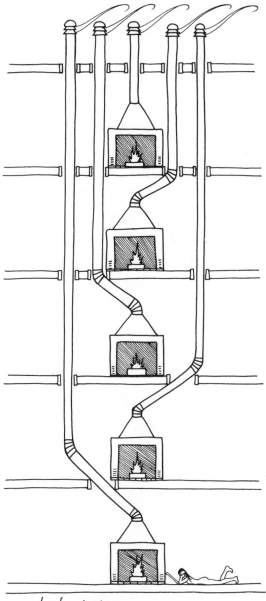

high-stacking
installation
of
air-circulating prefab
fireplace units

an air-circulating unit
for an existing
fireplace

hot air
out

cold air in

absorbing the warmth and radiating it back out—most of it back at the coals to keep them hot and the rest at you and me to keep us warm. A sheet metal firebox, particularly one with air circulating around to steal its warmth, can't do either as well. A prefab metal firebox of standard design is to a brick Rumford what a tinny little sheet-metal stove is to a big Austrian Styria stove of cast iron and firebrick. No comparison.

Of late, several innovations have cropped up that do make the circulators somewhat more effective heaters. Several inventive people have installed blower systems—some piping warm air throughout an entire house. With good insulation, such systems could keep you warm, though on cold nights it might be necessary to add wood in the wee hours.

And I've spent several pleasant weeks in the cool New England fall in a home where a thirty-year-old home-built system that forced air around a custom-made boiler plate firebox was all

**convection fireplace units**

the heat needed—though mornings come October were chilly. After a generation, the box, damper system and air conduits (of sheet metal) were beginning to rust out.

I'd place the steel firebox with or without an air circulating system in the category of occasional-use heaters. For serious fireplace wood heat, stick with brick. And even a good Rumford will fall apart in time. Maybe your great-grandchildren will have to rebuild yours.

the EFEL Fireplace

# CHAPTER FIVE
# Cooking with Wood Heat

Cooking with wood is cooking with skill borne only of experience. I guarantee that the first time you try baking bread with wood heat the loaves will come out as our first efforts did—risen at one end, shrunk at the other, burnt all around on the outside and half raw in the middle. Each fireplace, stove and kitchen range is an individual, its performance a function of the flue's size and draft, the type and quality of wood, time of day and season of the year along with the age and condition of the firebox linings, the length of time since ashes were raked out, and the effect of which design you picked from the infinite variety of draft, grate, damper and ornamental designs that have cropped up over the last 1,000 or so years of cooking appliance development.

In addition, cooking with wood is work; you can't set the timer or some gadget that turns off the heat when your roast reaches the proper temperature and go off shopping. The difference between wood and more modern gas or electric appliances is technical. But the difference between the cooks using each is in the mind and spirit. You can let the modern stove do the job by regulating the length and intensity of heat output on the burner or in the cooking box. With wood—particularly if the fire is keeping you warm in addition to cooking your dinner—the heat source is fixed. You've a fire and that's that. To cook properly you must regulate the food, moving it to hotter or cooler places near the grate fire, on the stove top or whatever. Since the fire intensity will vary in output as new wood is added and air supply is changed, the moving job is a continuous one. Even in a good wood range where you can regulate the flow of hot air and smoke around the oven, you must constantly make adjustments.

Then there's the time that goes into feeding and poking the fire, hauling the ashes, cleaning the firebox and air channels and desooting the stovepipe. Wood cooking isn't for everyone at all times, and as the photos show, our own range has a set of gas burners, a gas oven, and a "salamander" (a broiler) on top. We use them plenty. Plus we have a little superinsulated electric oven made by Farberware that doesn't waste heat—keeps it all in just as the wood oven does. The bread gets baked in the electric appliance on a warm August morning you may be sure.

However, anytime the kitchen needs warming, the wood stove is fired up and we cook most everything on it. There is something about a wood-fire cooker and the time and care that go into caring for it that also demands that you put more time and care into preparing the meal. Meals cooked over a wood fire just naturally have to be worthy of the time and effort that have gone into producing and maintaining the cooking heat. You don't build a fire to warm up a TV dinner. And, wood heat seems almost to demand natural foods—old-style *real* food without the chemicals that go into most stuff in the supermarket.

## The Open Flame

Cooking in a modern decorative fireplace, whether of steel or brick, is possible but not very practical. They simply don't retain and radiate the needed mild heat in a workable fashion. Fireboxes are too small, masonry too skimpy. To get a fire to generate enough heat to roast a haunch of meat, you must get it too hot—with a great roaring blaze that wastes fuel and gets so hot you can't stand to be near it. The decorative fireplace can't maintain a decent bed of coals long enough to keep a stew simmering properly either.

I'll assume that if you do cook over open flame, you have a properly large (Rumford or similar design) firebox or an open-front stove (Franklin of 34-inch or larger opening, or one similar). Either will keep a good bank of ashes and a low fire to radiate the gentle heat you want for cooking. Admittedly, the

Cooking in the old way starts with the proper fireplace. Gay Pietz lives in a circa 1765 firehouse in central Massachusetts that still has the original kitchen hearth. Here she stirs the beans in a pot on a two-trammel rig under the cooking crane. Since it's spring, the andirons are positioned so the horizontal log rests are close together to support a "short fire," which produces a low heat that's fine for cooking but not for heating. In winter, the andirons are switched so both the ash bank and the logs extend almost the entire width of the firebox, producing a fire that comfortably heats the center of the Pietz house, even on the coldest nights.

fireplace with a shallow firebox is a better open-flame cooker. It has a large enough opening to provide a greater variety of heats and more cooking area of easier access to the cook than any wood stove I've ever seen. But we've done up many a stew in and on the open-front stoves we've had and watched a good many pounds of meat roast to perfection in front of them.

Whatever open fire you have, it is important to get the firebox well heated, a good bed of coals banked, and the wood gently burning down. If you are keeping a permanent fire as described earlier, you've no problem. If starting cold, light up early in the morning for the dinner meal—preferably the morning of the day before. At first you may want to test the coal bed and heats in various areas of the fire. We test fires out with a little oven thermometer that you can buy in any hardware store for a dollar or so. Usually the old-fashioned roasting heat of about 350 degrees F. will be just a few inches out in front of the opening. A good broiling temperature of 500 degrees F. and more exists right over the coals.

### Using the Proper Utensils

Cooking over an open fire is the oldest form of the art, and you'll surely want to get in a supply of old-time cooking utensils. No need to spend a fortune for real antiques. With today's growing interest in wood heat, more and more firms are making such gadgets as trammels and trivets, cranes and danglespits. Even

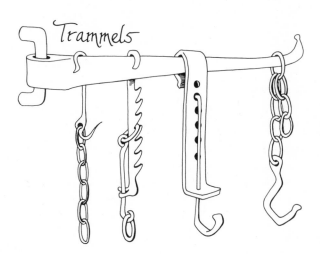

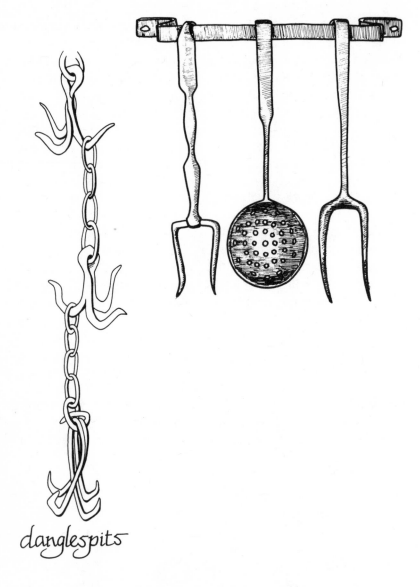

danglespits

the local hardware store sells pot hooks and bean pots nowadays. We've adapted campfire cooking equipment to inside use and many items can be made from readily available modern materials. A blacksmith (if you can find one—more and more are going back to work nowadays) would be delighted to work you up a big cooking crane. If need be, a welding shop can put one together for you.

### The Crane

The swinging crane with a cast-iron pot hanging from it is probably the most recognizable accessory. Most modern Franklin stoves have them available, and these can be adapted to most fireplaces as well. They are triangular affairs that swing out over the fire. At the end of the top leg there is a hook so the pot won't slip off when you swing it out over the fire. In small stoves and any little decorative fireplace, the crane and a small pot will likely fill one whole side of the firebox, and you will have to move coals around and do a lot of pot swinging to keep the heat correct.

In a big walk-in or Rumford-design fireplace, however, you've got the proper amount of space to set up a real colonial cooking hearth. Cranes were usually located at the back and left side of the large walk-ins and were of a length to swing out and have the hook about in the center of the fire bed. They were also placed high up in the firebox. Very large pots could be hung directly on the hook of the big back-wall crane. But for smaller containers, there was a variety of ingenious raising and lowering devices, called trammels. Roasting spits were supported on link

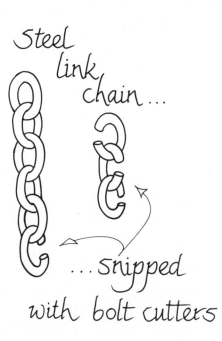

Steel
link
chain...

...snipped
with bolt cutters

chain; height was adjusted by hanging the spit and roast on whichever link provided the best cooking heat. Pot hooks were adjusted the same way.

A new or serviceable old wrought-iron trammel would probably be hard to come by. But you can easily get a length of steel link chain. Just use some stout bolt cutters to snip the middle out of one side of the lower link. It will serve as your pot hook. You can leave the rest of the chain as is and adjust the height by raising and lowering the whole device. Or cut the sides out of all the links and have individual C hooks that can be added one at a time.

*Installing a Crane*

Putting a crane into a steel-sided fireplace (if you must) is simple if you are putting the whole thing in new. Just get a crane made for the unit, drill holes and bolt it on before you frame the fireplace in. You can do the same with cast-iron stoves. You'll need high-speed carbon steel drill bits and a good power drill for either job.

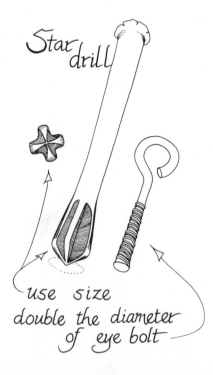

Star drill

use size double the diameter of eye bolt

Installing a crane in a masonry firebox is more of a job. First, get a star drill of the proper size (twice the diameter of the bolts or pins you will use to attach the crane). If using a manufactured crane, get bolts that will go through the holes in the frame, at least an inch-and-a-half long. If using an old crane or a custom-made one, get eyebolts to fit. One possible design of a crane welded from angle iron is illustrated.

Mark your holes—in mortar joints between bricks is best and if making your own design, arrange for bolts to meet with joints rather than bricks. Hammer away at the star drill, turning it a bit after each blow. In time you'll wear a hole in the masonry. If drilling into soapstone, use a power drill with a car-

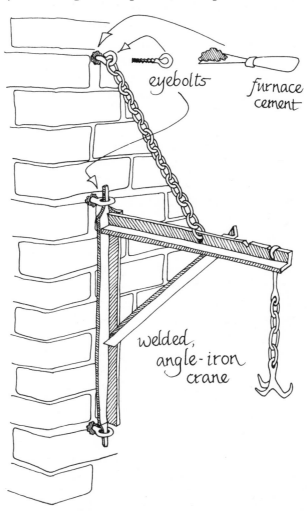

eyebolts

furnace cement

welded, angle-iron crane

borundum masonry bit as the stone is too fragile to hold up to a star drill unsupported. You can use the same rig on bricks and mortar, too, needless to say.

Now turn the bolts in, with stove cement packed into the holes around them. Don't use the lead or plastic expanding plugs for putting bolts into this masonry, as a hot fire will melt them out. In the three-bolt arrangement illustrated, you would want to put the bottom two bolts in, let cement dry, then put on the crane before attaching the top bolt. If you're using the upper supporting chain in our design (really necessary only in weak brickwork or for really big cranes to carry cauldrons), put this in last, as you want the chain to be taut enough to provide support and you won't know where the topmost eye bolt should go till the crane is up.

*Cast-Iron Cookware*

The only food containers that look right over a wood fire are of cast iron. And fortunately, many foundries are again making ironware, some of them from the same molds that turned out the pots, skillets and Dutch ovens our great-grandmothers cooked with. For cooking and baking under your crane, get several sizes of iron cooking pots with iron lids. Some are sold with Pyrex glass lids, which let you see the cooking but which never seem to last long around our place. Make sure the pots have stout steel handles (bails). The best selection of ironware I know of is at the Cumberland General Store, Route 3 Box 479, Crossville, TN 38855. For three dollars, they sell a mail-order catalog containing over 250 pages of tools for self-sufficient living. Their instructions on how to properly season ironware are the best I've seen in print, so I'll take the liberty of reprinting them: "Always rub your new cast-iron utensils, inside and out, with animal fat (lard for instance) or vegetable oil. Put in HOT oven and heat. HOT! Do this several times. (You might want to do this while doing other baking, so as not to waste fuel.) When you get your utensil 'cultured and cured' properly, it'll put Teflon in the shade!!"

*Seasoning Cookware*

In a fireplace, of course, you season by putting the liberally greased ironware on the banked ashes. The purpose of the seasoning is to open up the minute pores in the iron, let hot fat in, then burn it on to form a glaze that acts as a barrier between food and the iron. Never scour an iron pot; if food sticks, it's better to toss the pot on the coals or bake it in the oven to burn the food to a char. Should the pot rust anywhere or should water-cooked foods have an iron-y taste, repeat the seasoning process. Seasoning is smoky and smelly, but essential.

In time a black coating will build up on both the inside and the outside of well-used iron pots. On the inside the coating will be thin, hardly perceptible in the middle and lower pot. At the inner top and all down the sides, it will be thicker. Nothing but well-charred fats and oils, perfectly sanitary, and as the folks from Crossville say, it'll put Teflon in the shade.

By the way, if you've tried ironware on modern stove tops and ovens and found it unable to take and hold a good seasoning, don't feel bad. So have we. Part of the problem is that with a gas or electric burner or bottom-fired oven, the iron bottom gets too hot too quickly, throwing the seasoning into a water-based dish. Cook slowly. Louise also suspects that acidic foods remove seasoning, so stews containing tomatoes go into ceramic pots.

You should never have problems of losing seasoning over an open wood fire. The gentle, overall radiation from a good fireplace will heat the utensil slowly and thoroughly. In a good

hearth
pot-oven

lid filled
with
hot coals

firebox, the sides of a pot will simmer almost as much as the bottom. What distance to hang a pot from the fire is something you can learn only through experience with your own equipment. A candy or deep-fat thermometer is probably a good investment at first, particularly if you want to use the marvelous capability of ironware to keep an even heat for deep-fat cooking of French fries and the like. In short order, though, you'll be able to do wonders unaided with the crane, your ironware and the fire.

I'll not offend you by telling you in great detail that you fry bacon in a skillet, make stews in a stew pot, cook chicken in a chicken fryer and fish in a fish poacher, should you want to buy all the appliances. Or that you can do it all in a single flat-bottomed skillet. Get what you want to cook what you want. Pots with bails are about all that work with the crane. Handled skillets and the like need some bottom support, and I prefer to equip a fireplace with two or three log rests. You can scrape coals as wanted out to the front of the firebox, set skillet or pot on the log rests and let it cook away. The log rests are cast iron and effectively transfer additional cooking heat from the fire banked at the back of the firebox. In most stoves, you are restricted to the smallest sizes of pots and skillets; in a big fireplace, you can brew up some really big stews.

## Baking, Roasting, Broiling over an Open Flame

I suppose that a Dutch oven is so called because the Dutch bake in it. Louise isn't Dutch, but she turned out plenty of round loaves back in the big walk-in fireplace we used to have in Pennsylvania. For the novice an oven thermometer is a must. Put a thermometer inside your biggest Dutch oven on the crane and move them up and down, in and out through several day's woodburning. You'll soon find out what combination of pot positions and fire produce and maintain your preferred baking temperature. Louise always used two big six-quart ovens. Loaves were let rise to the proper size in one placed out in front of the flames, where it was warm but not cooking hot. Then she'd bake the first while another was rising for the final time. This alternating of ovens would go on as long as she needed.

With a deep walk-in fireplace, the bricks all around radiate enough heat to keep a Dutch oven quite evenly heated, though the pot and bread usually got turned every fifteen minutes dur-

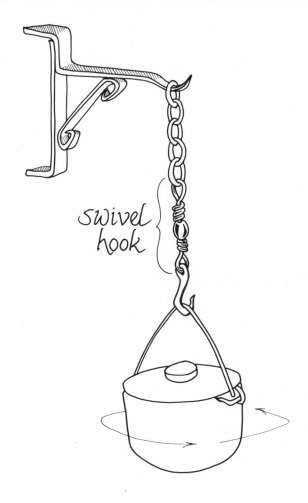

swivel
hook

ing cooking. With a shallow-boxed Rumford, all the heat radiates into the room—fine for heating, but not for baking. Cast iron retains heat, and you could probably arrange a swivel hook and keep turning it every minute or so. I can imagine a motor or spring-run gadget that would turn it for you. Easier is to erect a reflector—like a giant camp oven. (One of which you can certainly use for quick breads such as biscuits.) But for slow-baking yeast breads, it's best put up a regular reflector—make one of shiny sheet metal as illustrated. Again, trial and error alone can tell you what position is best for the Dutch oven and reflector. Here's Louise's favorite recipe for Dutch oven bread.

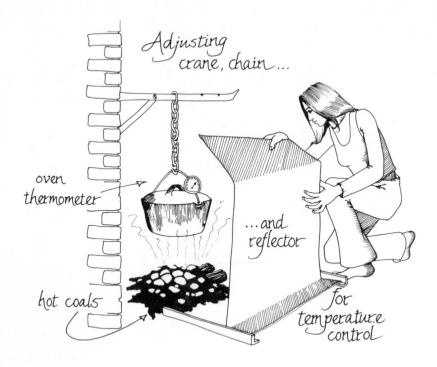

Adjusting crane, chain...

oven thermometer

...and reflector

hot coals

for temperature control

### Louise's Dutch Oven Bread

Arrange fire, pot and reflector to maintain a good hot 400 degrees F. for an hour per loaf. We found that this heat in our fireplace was in the corner about a foot from the back and left side of the brickwork, the pot a bit over a foot above new, hot coals under ash several inches deep. The coals had to be renewed from the heart of the fire every quarter or half hour, and the reflector placed from one to two feet away. A small oven thermometer was cinched over the bail on the side away from the fire. Checks every few minutes tell if the heat is in the 400 degree F. area. Adjusting the reflector out gives you 20 degrees F. or so of cooling control. A small shovel of red coals pressed under the ash and the reflector moved in gives somewhat more of an increase in heat. You may want to use a camp or hearth pot oven with legs and a lip around the lid top to hold coals to brown the top of the bread.

For each large, double-sized round loaf, mix one package of dry yeast in warm water containing a few pinches of light brown sugar. Scald two cups of whole milk (bringing it to not quite boil-ing) and while it is still hot, add two tablespoons of light brown sugar, two teaspoons of salt and one tablespoon of sweet (unsalted) butter. Mix well, let cool to lukewarm, then add the yeast, stir again and combine with five cups of unbleached white flour and a cup of stone-ground whole wheat flour. Here you can also add up to a half cup of your favorite grain supplement—wheat germ or any variety of seeds or nuts. Stir till moderately firmed up, then turn onto a floured surface and work in an addi-tional half to three-quarters cup of white flour as you knead to work the gluten out. Knead until the surface of the dough is smooth and has a satin shine to it, but no more than ten minutes. Lightly butter the inside of the pot and the top of the Dutch oven. Form the dough into a ball and put in the pot. Cover the pot with a wet towel, and let the dough rise out in front of the fire for up to two hours or till it has doubled in size. Now punch the ball down lightly, turn it over in the pot and let it rise and double again, up to another hour. Punch it down lightly once more, turn in the pot and let it get its wind for a quarter-hour. Then put on the top slightly ajar and put the oven on the hook. (If not using a reflector, leave the top off.)

The oven must heat up so the inside of the bread has reached the correct temperature, then it must cook with the top off for an hour or more (or less). Only experience will tell you how long this will take in your own kitchen. It varies with size and thickness of the pots, with the shape and heat of the firebox, the reflector and much else. But we found that if the bread is burnt outside and raw inside, the heat was too high and not long enough. If burned on the bottom only, the pot was too low or the fire not well banked enough. If burned on the top and not the bottom, the reflector was too close. That was in our fire-place. You may have better success by dividing the recipe in two and cooking two smaller loaves in smaller ovens. Perhaps you will find it best to preheat the pot and after the quarter-hour rest, pop the dough into the already cooking-hot oven. Be sure to leave the top on but ajar until the bread is fully rerisen. That way the bread can heat evenly for the in-oven rising period but gases released by the yeast can escape. You'll simply have to ex-periment, and be prepared for some initial disasters if your luck is as bad as ours was.

## Cooking Meat

Baking bread in a Dutch oven takes a Louise-like patience—willingness to suffer through a lot of burnt crusts. But the fool-proof glory of open hearth cooking, that even yours truly can't ruin, is grilled or spit-roasted meat. If you char the outside of a steak, it just adds to the flavor. Or if your slowly turning chicken or pork roast looks as though it will brown before it is cooked through, you just move the roasting spit farther from the fire.

The term to *grill* meat comes from the utensil used in the old days to cook thin cuts of tender meat: a sheet of metal with slits cut in it in the shape of a grate. *Gridiron* is another name for it, a word now mainly used to describe a football field, with its al-ternating stripes and open patches of grass. A proper gridiron is made of cast iron, and the only ones being made these days that I know of come on hibachi pots—the little Japanese-style char-coal cookers. Wire grates or such used on home barbecues would work for a time, but would warp if used the way the gridiron should be. We bought an extra foot-square hibachi pot grill that fits just right between the log rests in the stove.

To grill, I simply rake hot coals to the front of the fire, put the grill on the log support till it has heated to smoking hot.

*gridiron*

*spout for pouring grease from trough*

*pivot grills
and
cranes*

Then the steak goes on the grill and is left over the hot coals till the part of the meat surface touching metal is burned black and the open part is well-seared. Then, before fat can begin falling onto the coals and catch fire, I cover the fire with warm ashes. The steak is cooked five to ten minutes to a side, and I'll tell you, no backyard barbecue over charcoal briquettes can hold a candle to the flavor. A few splinters of hickory wood thrown on just before the coals are covered will add a genuine hickory-smoke flavor. (Be sure not to use any evergreen or birch wood for any part of the broiling fire, as the aromatics can add an unpleasant cast to the taste.)

You can finish the dinner by baking potatoes—whether wrapped in foil or not—for about an hour in the cooler part of the ash bank and cooking up your vegetable in the bean pot on the crane.

*The Roasting Spit*

Now, the most spectacular aspect of cooking with wood heat—the roasting spit. You can still find rotating spits before the huge fireplaces in English manor houses and their American counterparts. The supports are huge affairs, often capable of holding a dozen or more horizontal spits before the flames, and the spits are turned with complicated chain-and-gear or rope-and-pulley devices operated by servants. These spits could hold entire steers for a day or two of cooking, as well as enough fowl to feed dozens of people at once. Few of us can afford to live on that scale anymore, and modern roasting spits are geared to smallish roasts and a chicken or two. My favorite modern spit, though, is a true Roasting Jack designed by David Howard to complement the Rumford fireplaces he builds in his post-and-beam houses. A big one could hold a half-grown hog. It operates by gravity; you crank a more-than-two hundred-pound weight up to the roof peak, and as it descends slowly, it turns the spit and the dinner on it through a pulley-and-clockwork arrangement.

Our own spits have been a bit less ingenious. I arranged a simple spit by banging two notches in the hollow metal spheres on the tops of a pair of old andirons. The bird or haunch of meat was skewered on a length of steel rod, which was suspended between the andirons. The reflector was pushed up and a drip pan put under the meat. It wasn't a revolving spit; the meat had to be turned frequently and the drippings ladled up to baste. But it worked.

As I'm sure you've found if you ever tried to operate a rotisserie of any sort, it is nearly impossible to get the spit through so the meat is equally balanced all around, like a wheel is on an axle. One side or another will always be heavier and tend to flop, as it comes over the top, often working the whole piece off the spit. You can hold it in place with cooking fork props. But the best remedy that we've come across is the meat holder found on modern rotisseries, a pair of two-pronged forks that ride along the spit and have thumbscrews that can be tightened so they'll stay fixed and keep the meat centered on the spit. There are models to go into kitchen ranges and outdoor cooking units, and they can be adapted to your open hearth.

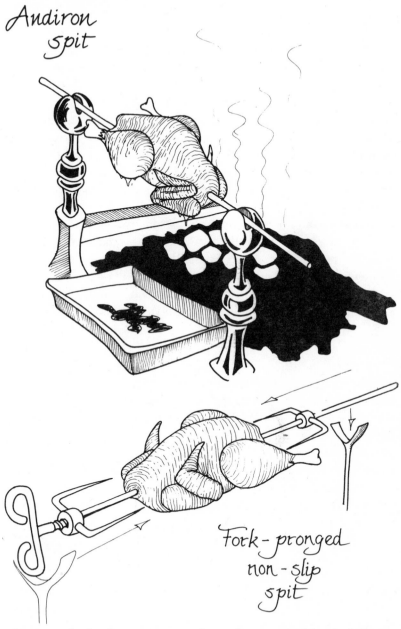

*Andiron spit*

*Fork-pronged non-slip spit*

Ours is a little electric-powered one from a neighbor's defunct outdoor cooker. I rigged up the old hood and motor over a dripping pan to sit in front of the fire. The hood helps reflect heat and makes a dandy cooker.

## Stove Top Cooking

The essence of stove top cookery can be pretty well summed up in one sentence: build a hot fire and move pots around on the stove top till you find the temperature you want. The steel or cast-iron top will be hottest just over the heart of the fire, in the center of most Franklin and other box-heating stoves, under the oven or firebox in a kitchen range. Most cooking stoves and many heaters have holes in the top fitted with lids that can be removed to expose a pot directly to the fire. Ranges usually have at least one three-part lid with a center circle and two surrounding rings so you can adjust for small pots. Our range, built in the waning (old) days of wood heating, has one three-parter and one solid lid on the left (over the fire) and three smaller ones to the right. When there is a good baking fire going, there will be live flame under all lids and the entire cook-top will be hot enough to boil water with the lids in place. Cooking over open lids leaves soot on your pots unless the fire is down to a bed of slow coals.

lid removed
for a
round-bottomed
pot...

...or a pot made
to fit

in the hole

So we remove lids only when there's something to be brought to a fast boil over a low fire or if we are using a round-bottomed pot or a skillet that is warped or dented and can't sit flat on the top.

**Using the Proper Utensils**

But no cook has enjoyed maximum flexibility in top-cooking till he or she has used a wood stove. The entire stove top is a

three-
part
lid...

...for
small pots

...and
smaller

cooking surface ranging from super-boiling-hot to just-warm temperatures. If using steel pots or skillets, you can literally see the level of boil change as you move a pot around the top. (Cast-iron cookware keeps its heat too long to be very good for dishes requiring instant changes in temperature.) The top is the only true self-cleaning cooking surface I know of. Any spills burn to carbon in short order and are just swept into the fire. If you like, you can fry directly on the iron top, though we prefer to use iron skillets. And the best hotcakes known to hungry man come from the big soapstone griddle that's let warm up on the heating fire overnight, then heated to hotcake temperature—so a drop of water just skitters—over the breakfast fire. You never use fat on soapstone; just mix up the favorite skillet bread recipe and bake right on the stone. It never sticks, never has to be washed and will last for eternity. (The stone will split in time, however, a natural reaction to repeated expansion and contraction. The metal ring around the stone is to hold it together when the expansion joint does appear, as well as to provide an anchor for the handles.) Several mail-order houses sell the griddles, or you can order one directly from the maker, The Vermont Soapstone Company, Perkinsville, VT 05151. The price is hefty—$30 and up a few years back, but if the stone is given proper care, you

Soapstone griddle

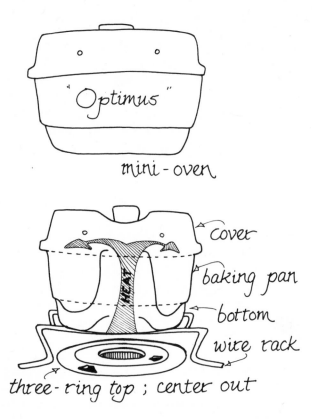

mini - oven

cover
baking pan
bottom
wire rack

three - ring top ; center out

and yours won't have to grease another griddle for the next several generations. You *can* grill meat on the stone, but then it will have to be washed, and grease is bound to run down under the copper ring and end up dripping on the floor, so don't. Once it splits, let it cool before you remove it from the stove. Hold by both handles. It is best to store soapstones flat, not hanging. Don't drop it!

Another stove-top bread baker we like for quick breads is a version of the Oriental volcano cooker. Ours is from Sweden, made by Optimus and sold by several camping equipment manufacturers. The diagram shows the principle. The cooker was designed for an open flame, but by removing a small cooking lid on the wood stove, you get an even better all-over heating effect. Breads turn out brown and crisp top and bottom.

A gadget that Louise finds useful is a trivet, a cast-iron plate, usually of fancy scrollwork design, set up on three legs. It helps prevent hot serving pots from burning the dining room table.

*especially good for (sponge cake baking*

But it is also useful, placed on the cook-top, for keeping dishes hotter that they would be up in the warming ovens, but not cooking hot. Another helper in regulating heat on the top is a set of coils. Just as the name implies, they are lengths of steel coil, like an old-time screen-door spring, twisted together at the ends and pulled out into as near a circle as you can get. Over a coil a pot will simmer slowly without the dish sticking or burning, as it might right on the hot cast iron. You'll get more heat to the pot with a snail, a longer spring turned into a flat spiral and held flat by two lengths of stiff wire running through the coil in an X-shape. None of these devices are sold anymore, but they are easy to make from door springs and clothes hanger wire.

*trivets*

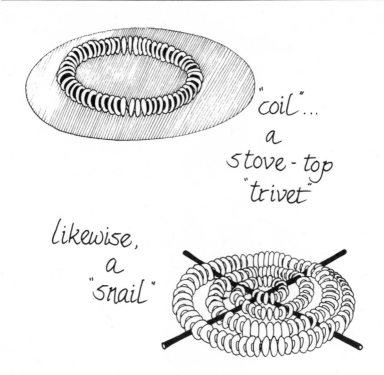

"coil"...
a
stove - top
"trivet"

Likewise,
a
"snail"

## The Wary Cook

A few cautions are in order for the novice stove top wood cooker. First, there is no flame or glowing element to tell you the stove top is hot, and it will retain enough heat to burn you even if the fire is nearly out. I can tell you that the first time I put a palm on a unexpectedly hot stove top was my last. I hope you have more sense than to try it even that one time. Fortunately, if the top is searing hot—capable of burning you seriously, it will radiate heat far enough to warn you. But we never let little kids roughhouse anywhere near a hot stove and whenever they insist on helping with the meal preparation, they wear sturdy potholder gloves.

Also, *all* of whatever utensil you're using on the top will heat up, including wood or plastic handles of the sheet metal skillets, for example, that would stay cool over a gas flame. You can singe a hand on one. We've found that Pyrex glass pots aren't best for wood stoves. A modern sheet-metal range is tinny enough that a

coffeepot that is dropped will bounce enough to stay in one piece. Not so on iron ranges. Also, we've had the pump stem of a glass percolator break inside the pot when it was empty and let stay on the hot surface. It just got too hot, I guess. No matter, because wood-cooked coffee tastes best when the grounds are mixed with a pinch of salt, an eggshell, and water from the big iron coffeepot that sits on the stove back—humming and humidifying the air all winter—then boiled up, let sit long enough to settle, and finally enjoyed hugely, muddy color and all.

## Queen of the Kitchen—The Wood Range

A lot of wood heat books I've read rave (and it's perfectly justifiable) about the glories of a woodburning kitchen range and unmatched quality of breads and cakes baked in their wonderful ovens, and we agree with them all. (Hot air is not continually run through woodburning ranges, as it is through gas or electric

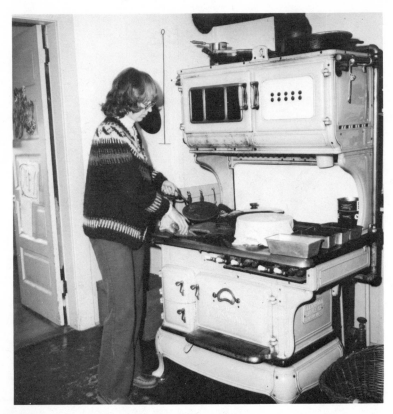

There is no place on earth that's better for raising bread than a wood stove. The place you put the rising pans varies with what is going on in the oven or on the cooktop. Here, it's near time to begin baking, so the oven is being heated up. So, the loaves go down to the cool end, so as not to rise too fast and get all cheese-holed.

appliances. The wood oven is closed tight; hot flame and smoke are channeled around it—top, bottom and all sides but the front in the best designs—so the heat radiates from all directions. There is no way to bake a better loaf of bread. Indeed, once you've tasted your first success with the range, you'll wonder how the home appliance industry managed to foist their gas burners off on our mothers.)

However, the how-to of range use is hard to find; not only were the modern authorities of little help in teaching Louise and me to operate a wood range, even the 1925 booklet *The Secret of Better Baking* by Mary D. Chambers, (once reprinted by and perhaps still available for small change from the Portland Stove

Foundry Co., Portland, ME) fails to divulge all the secrets. It assumes the use of coal, and it just naturally expects the reader to know a lot of information that has largely been forgotten in the past two generations. For example, the booklet assumes that we all know what a spider is, as well as the difference between a dock ash grate, a plain grate and a triangular grate. I still don't understand grates to be honest. The dictionary says a spider is a long-handled skillet, originally with legs to go over coals. And here is what Louise and I have picked up from Ms. Chambers and other genuine experts, along with a lot we learned the hard way.

### Shopping for a Range

Back at the beginning of wood heat's rebirth, the dream of every would-be wood cook was a Princess range. Featured in every country living magazine and catalog, they were circa

*old range*

1880s, black-iron and nickel-plate beauties, handmade by the Portland Stove Foundry from original patterns. In a matter of a year or two, price went from about $500 to better than twice that, and the waiting time for delivery to well over a year. But it was worth it to many for the only available modern reproduction of a classic woodburning range. With the wood-heating boom, however, the money men jumped into the few old foundries to hold out against Standard Oil and friends. At this writing, Portland has rolled over to the airtight-door, baffle-plated firebox fad, has retrofitted a couple of good, old, simple designs and retired the Princess and entourage as unprofitable. R.I.P.

A few old-style kitchen ranges are beginning to trickle in from Europe. We've seen several very primitive range designs from the Scandinavian stove makers and one or two approaching 1890s quality from Ireland or central Europe. Ads are beginning to pop up now for other domestic-made iron stoves of old designs, and I have little doubt that time will prove some of them to be excellent. However, the surviving old, established firms pretty much limit range production to white enamel, modern box designs, most of which are mainly coal stoves. A few such are also imported—all good stoves with adequate instructions on firing, if not on baking, if you like an antiseptic '30s design in your woodburning range. The Far Eastern hustlers are into ranges too—lightweight, tacky travesties of antique designs, of a quality that will make you ill if your wood stove knowhow extends much beyond an admiring inspection of the Tolla airtight that warms your neighbor's study.

At least one new company that I like the looks of is putting out steel ranges with cast-iron fireboxes in the larger, serioususe models. This idea makes a lot of sense for many folks. Prices are right, units lighter and smaller than all-iron ranges of comparable quality. Still, these are new designs, and just a few of the dozens that are sure to come along. By all means, check them carefully. In no modern design made will you yet find the sophistication of design and operation, the sheer beauty and class, my friend, of a genuine old-time woodburning range. And for my money, it is more than worth your time to search one out for yourself—*very* carefully. The problems of evaluating old heating stoves are compounded when you go looking for a kitchen range. Since hundreds of manufacturers turned out

*modern range*

thousands of models, I doubt that there is anyone living who could list everything you should look for in judging an old stove. I know I can't. But if we refer to the cutaway drawing to discuss how the range works, you'll have a fair idea of what to look for.

*The Boot*

Very first thing, go around in the back. If the stove does not vent directly through the back of the cook top, there must be a cast-iron elbow-like projection, the boot, attached where smoke is emitted and forming a 90 degree upward turn, ending in an oval form to fit stovepipe. This is the part most often missing in the old stoves I've seen. You *must* have one. Even if you managed to adapt a piece of stovepipe elbow to serve as a makeshift boot, the hot gases roaring straight out against the turn in the pipe when you are baking biscuits would burn it through in no time. You could conceivably have a new boot cast, but that would require someone to make wooden or plaster patterns, two of them to be welded or bolted together, and you'd be better off waiting for a new range. The boot may have a damper built in, most likely attached to an operating lever that comes out at the back of the range. This is a convenient thing to have,

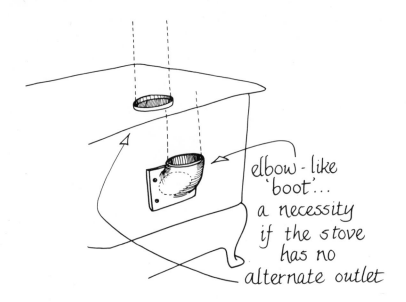

elbow - like
'boot'...
a necessity
if the stove
has no
alternate outlet

but if it is all that is missing on an otherwise good range, you can install a regular stove damper in the pipe, perhaps running the operating rod of the damper out through a hole drilled in the sheet metal back that supports the top shelf or warming ovens. Point out the missing part and offer the seller half of what you think he really wants for the range.

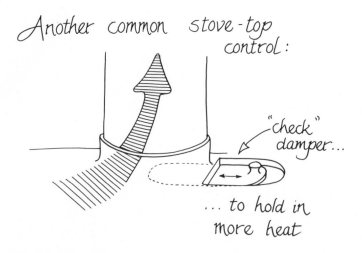

Another common stove-top
control:

"check"
damper...

... to hold in
more heat

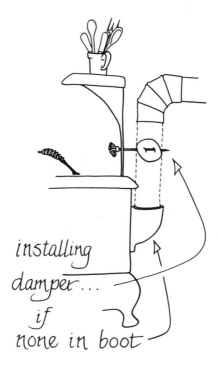

*installing damper . . . if none in boot*

### The Firebox

Now check out the firebox on the left side of the range. There will be a draft control at the bottom, another at the top, both located at the side or front (or both), and perhaps the top one hinges out (for broiling). Make sure all hinge pins work, that the draft slides and the doors open and close nearly airtight. Don't buy any with damage you aren't sure you can repair with whatever skills you possess or can hire. The grates halfway down in the firebox side—a grid or a metal rack or a plate over a set of rods that can be shaken or rotated—ideally will be sound and operable. You'll find a firebrick lining all around if the stove was made for coal and cast-iron plates, perhaps removable, if it was meant for wood only. Ours has a double set of liners, one permanent, another of thick iron plates that can be removed. This is a feature of some wood/coal stoves; the extra liner is meant to come out when using wood. We leave ours in to keep the old girl going at least another half-century. Many will lack liners and grates, having been assembled in the 1920s for

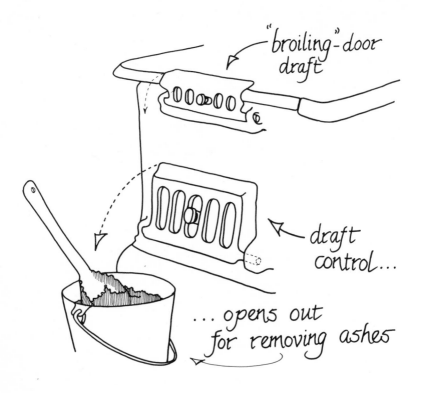

"broiling"-door draft

draft control...

...opens out for removing ashes

kerosene. You should be able to improvise grates—or see if the stove owner can do it for you.

If the grate shakers don't work, it is no tragedy for wood use. You can just push ashes down through the grates with the poker. Revolving or shaking grates are necessary with coal, as the ash and klinkers have to be removed each morning, or else the grates will burn out. It's not the case with cooler-burning wood. If you find a stove with a firebrick liner though, be sure that the liner is not burned through to the metal, or else be prepared to patch it or reline with new firebrick or furnace cement. The firebox walls are the most likely part of the range to be cracked or warped. No matter what you're told, no one can weld a cracked firebox so it will hold up for a long time, even if he does use nickel welding rods. And the firebox of a cookstove probably gets more stress due to changing heats than any part of any solid-fuel heating device. So check the firebox particularly well. If it can be relined with sheet metal or firebrick, a small crack in the firebox wall doesn't condemn the range to the dump. Or you can

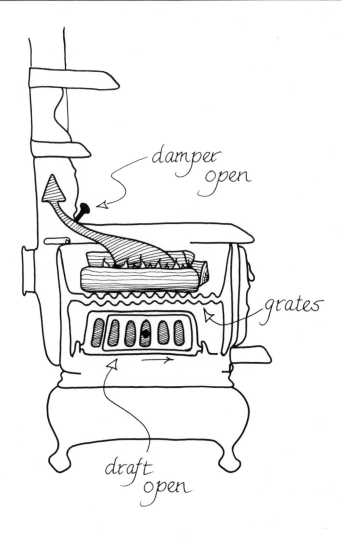

damper open

grates

draft open

*firebox side*

just plan on patching it with stove cement every so often. No big chore. A wood fire gets hot, but not nearly as hot as the coal fires that probably cracked your stove. Warped plates along the firebox sides, front or back should be carefully inspected. Here, as elsewhere on the stove, if you see a lot of pits, the stove has been allowed to rust and most probably it's been out in the weather for years. Triple check hinges, et cetera.

### Is It Repairable?

The older (and costlier) the stove, the more primitive was the casting technology of the time and the more likely you are to have warped plates. For really serious daily use, I just wouldn't buy any genuine antique, warped or not. Any stove manufactured much earlier than 1860 belongs in a museum, I'd say. Too, I feel that most stoves built prior to 1900 just shouldn't be asked to go back to daily work. A lot of folks will disagree, including

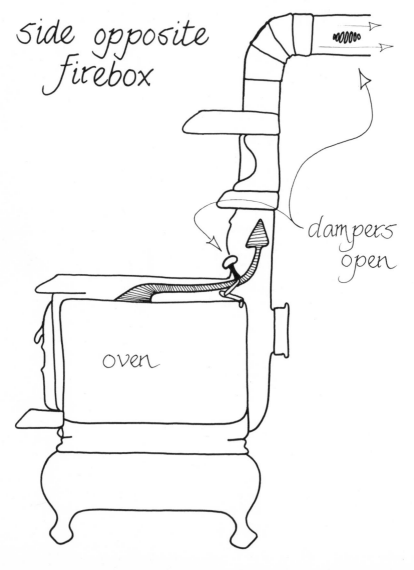

side opposite firebox

dampers open

oven

Richard "Stove Black" Richardson who runs the Good Time Stove Company on Route 9 in Williamsburg, Mass., and who knows more about old ranges and most other kinds of old stoves than anyone I know of, myself included by a long shot. Or maybe I should say that I'd not buy an antique range for serious use unless Richard or another expert like him surveyed it and gave it an okay. There are just too many models, styles and casting techniques that have come up over the years for anyone but

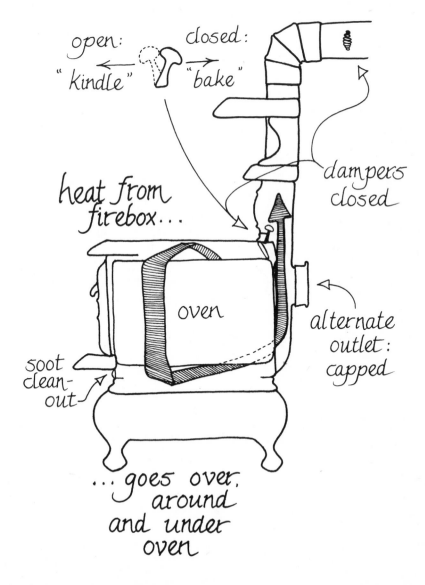

a full-time specialist to understand. And, unlike most heating stoves, ranges have several draft controls, integral dampers and baffles that the average person can't understand without a lot of trial and error, much less judge at first look.

However, there will be a lot of folks who just can't resist one of the old black-iron beauties. Well, in an antique don't be put off by rusted or even burnt-out grates. You can replace grates for wood very easily. Coal, no. But a good welder can cut you heat-proof steel rods, fine for wood heat, to fit in back to front or side to side, put in through holes bored in the firebox wall or on legs that go down into the ash pit or any one of a hundred improvisations. Forget the grates, but do worry about the firebox plates. Put back into use after two generations out in the weather, a warped and cracked stove could literally fall to pieces on a first firing. If the removable ash box below the grate is gone, you can have a tinsmith make up another, or you can bend and rivet your own from good sheet metal. Don't let the ashes just collect in the lower chamber; the heat will (usually) burn out the sheet metal bottom in time. Perhaps you can simply line the ash pit with metal and a layer of sand as with a sheet metal heating stove. Check the stove top and lids for excessive wear or warping. You can probably purchase new lids from one of today's stove foundries as sizes were and are pretty standard. The holes where the lid lifters fit are nearly worn away on a couple of our own 50-year-old lids. I figure that in another 50 years, we or our replacements will have to begin looking for replacements for the lids.

### The Oven Controls

Now, remove all lids and as much of the stove top as you can. Use a penlight and peer into as much of the ducting as you can. There will be a small door at the back of the firebox, normally operated by a lever located at the range back to one side of the smoke outlet. It has two settings. At *kindle,* the door is open so fumes go directly into the stovepipe. This is the setting used to start fires or for top cookery. *Bake* closes the door, so the heat is pulled over the oven, all around, under and back up into the flue. This control *must* work and all parts *must* be in excellent condition or your oven will never function easily. I suppose that you could rig up a replacement that could be stuffed into the opening through the lids, but I'd be of no help in designing it. However, the door and handle parts are usually relatively simple, are bolted on and could be recast or improvised.

The air channels over and around the stove will probably bear the residue of the last fires lit in them: a solid packing of fly ash. You should be able to see the top-oven air passage through the stove top. In any good range, the sides and back will be accessible through ports. If you think that you'd never be able to get an ash rake, brush or vacuum nozzle into the crannies around the oven, don't buy the stove.

Finally, check the iron and sheet metal parts for rust. Most of the better stoves had such parts as warming-oven bottoms made for easy replacement with the removal of a few bolts. You'll have to make or have made any replacements, but the job is easy for any job shop. If hinges on doors are broken or badly rusted, you should look elsewhere. Hinge *pins*, though—if separate steel rods—can usually be replaced if you can get the damaged originals out.

*Optional Equipment*

Such extras as gas burners, water fonts and the like take extra scrutiny. Since our subject is wood heat, I won't go into gas, except to admit once again that the gas side of our Glenwood is a great blessing come late June. If you are lucky enough to find a range with water heating capability, don't pass up a good cooking range if coils are clogged or the font rusted a bit. They can be easily repaired.

Most water fonts (or fronts) are nothing more than a metal box. If yours leaks, try to remove the whole cast-iron top, rather than just the hinged or removable dipping lid. This is to get at the liner, usually tinplate or zinc. It should lift right out; if it's really in bad shape, you can go at it with a crowbar. (Be careful of the relatively fragile cast-iron outer shell if there is one, though.) If the leak is small, get your Mapp-gas torch and braze it shut with the brass rod. Don't use lead solder, which would hold, but would give anyone drinking the water a case of slow lead poisoning. If the liner is shot or missing, write down the dimensions and have a sheet metal shop make you up a new one.

Sheet copper is good but expensive. I think I'd just have it made of galvanized steel (brazed again, not soldered with lead), then have the whole thing given a heavy inside coat of any non-toxic metal: nickel, tin, gold or silver. Silver'd tarnish, of course, and gold would cost a bunch (though not as much as you'd think) but would last forever. Imagine casually opening up your gold-plated water front to show any gotta-keep-up-with-the-Joneses neighbors.

If your range has a coil, one or another arrangement of piping to heat water by running it through oven or firebox, it will almost certainly be partly clogged or will get that way with minerals if your water is at all hard. In the old days they used reams to clean the straight sections and let vinegar sit in the curves to eat the clog out there. You can do the same (and/or can avoid the problem by setting up a rain barrel and heating only naturally distilled, pure water), or you can take the coil to any

hot
water
reservoir

good heating contractor. Lots of modern hot water furnaces have coils in them to heat the domestic water supply. Ours does, and the coil must be cleaned annually. They open it up and run acid through for a reasonable fee. You can take your coil to the shop and they'll do it even cheaper.

## Wood-Fired Ovens

Now, your wood range is home, you've got her cleaned out, polished and spruced up and want to cook a meal. Okay, have a can of stove cement handy. Installation and piping is the same as any stove. No need for a damper in the pipe unless there is none in the range smoke outlet.

We've discussed the damper (the "check") and the oven diverter ("kindle/bake"). The way they work and the instructions—the way "check," "bake" and all are conveyed on the stove—vary as much as the mechanisms. Some are words molded into the stove top or control handles, others are cryptic symbols or abbreviations. But you'll figure them out. Basically, you (may) have a damper and (surely) will have some sort of

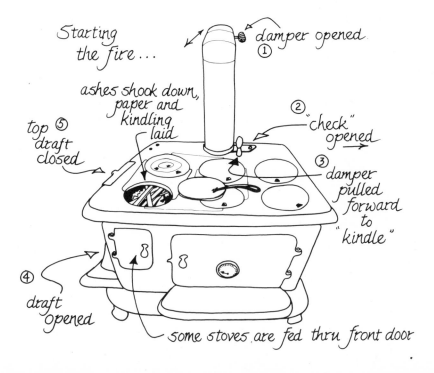

Starting the fire...

ashes shook down, paper and kindling laid

damper opened ①

② "check" opened

③ damper pulled forward to "kindle"

top ⑤ draft closed

④ draft opened

some stoves are fed thru front door

mechanism for directing hot gases into 1, the flue and, once the fire is going, 2, around the oven.

To make a fire: Open damper or "check" control fully, put diverter on "kindle," open the bottom draft control fully and open fire door or remove stoking lid. Let the flue warm a bit, meanwhile, holding a burning match down in the firebox. If the smoke and flame of the match are sucked out with vigor, you're halfway there. If not, there's an obstruction somewhere between stove and chimney cap that you'll have to find and cure. Lay a regular stove fire of crumpled paper, splinters and kindling. Drop a match in, and once the paper blazes, close the firing opening. Once the kindling catches, add more wood. In a few minutes, switch from "kindle" to *"bake"* and stand back.

If the stove is clean clear through, you'll hear nothing but a slighly reduced draft pull. If the whole machine begins to smoke, you have clogged channels in the oven and a session of excavating is in order, if you haven't cleaned the oven channels already. The clog must come out, and fortunately, these old ladies are not all that mysterious. You may have to drill out a few soft iron stove bolts to open up the cleanouts, but the technology is straightforward and once a good old stove is set up and cleaned, it should be good for the rest of your life and mine, as long as it's kept cleaned out.

I can't begin to tell you how or where the fly ash and soot that's clogging your range may be. Each model has its own series of smoke channels. You'll have to remove all top plates, bolted or spring-clipped-on ports at top, bottom, ends or back and dig away. The T-shaped ash rake we mentioned earlier is an essential tool for getting into the crannies. A stove brush (like a bottle brush, but on a long handle) helps and is good for cleaning out stovepipe too. I found that a vacuum cleaner with a long flexible pipe attached was almost essential to get into some of the harder corners. The smoke channels of all wood-fired ranges collect flying ashes and soot and can clog in as little as a month's hard use. Once you have yours cleaned of any extensive accumulation, plan a monthly clean out and the hard job will never reoccur.

If smoke spouts from any cracks in the stove (except from cook lids when draft is insufficient), remind yourself to caulk them with furnace cement next time the fire dies.

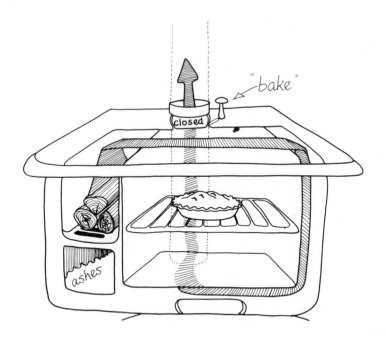

"bake"

closed

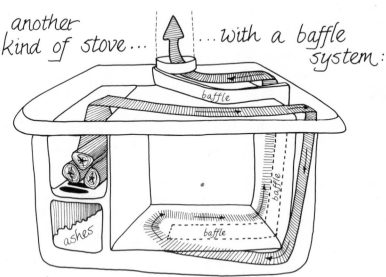

*another kind of stove…* …*with a baffle system:*

baffle

baffle

baffle

baffle

ashes

*heat goes across top,*
*down one side,*
*under ; back up the side,*
*across top again – and out.*

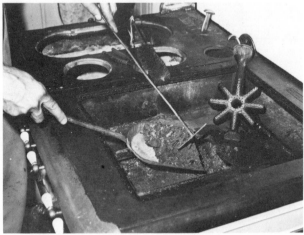

Cleaning a wood range is a messy job that invariably gets put off longer than it should. In addition to scrubbing the spots common to all stoves, you must remove the accumulated soot from the channels surrounding the oven. The ash rake and shovel are specifically designed to get into all the corners of the smoke channels, but it usually does take some probing around. A quick once-over each month and the fly ash never accumulates. Inevitably though, the newly acquired antique (like this) usually needs more than a quick once-over that first month.

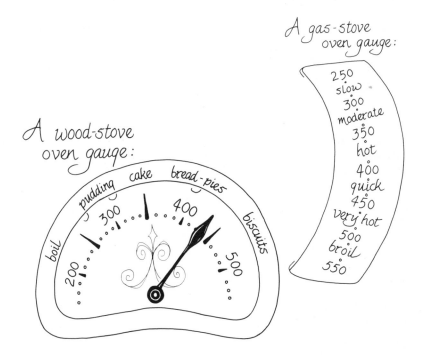

A gas-stove oven gauge:

250 slow
300 moderate
350 hot
400 quick
450 very hot
500 broil
550

A wood-stove oven gauge:

pudding  cake  bread-pies
boil  200  300  400  500  biscuits

## Gauging the Heat

Most modern wood-fired ranges built since 1900 or so have a temperature gauge in the door. It is *utterly* unreliable except as a very general guide to how hot the door is. Moreover, the old-time cooks didn't insist that recipes be cooked at precise temperatures. There were general temperature ranges, and the two gauges on our 1925 range were intended to help guide older cooks into the "new and modern" age of gas, by mixing the old terms with modern temperature numbers. But they serve as well to guide us newer cooks back into the old terminology.

How to attain the oven heat you want and hold it is information each cook simply must learn by trial and error. Louise finds that two small oven thermometers, one placed at the left front, the other at right rear on the oven rack, give the best reading both of general temperature and evenness of the heat. (And wood stoves never heat evenly, thus the accuracy of Jimmie Rodgers' 1920s era song line: *"I can smell yo' bread a-burnin'; turn yo' damper down. If you ain't got a damper, good gal, turn yo' bread around."*)

_biscuits..._

_...baked
with biscuit - wood_

_biscuit - wood...
    bone-dry
poplar for summer use:
        fast, hot.
    and lasts
just long enough
    for a batch to bake
alder for winter use_

## Controlling the Heat

Just as with any wood stove, you use dampers and draft controls, size and variety of wood to vary the intensity of heat. The more air passing through, the more oxygen is presented to the wood, and the faster and hotter it will burn. Generally, the smaller the split, the dryer and lighter the wood, the hotter the fire. *Biscuit wood* is used to attain the brief, but very hot fire needed for any soda or baking-powder bread—as the gauge indicates. Small splits of any dry wood will work. Small "trash trees" provide the best biscuit wood, if well-dried. A firebox full of poplar or aspen will last just long enough for a batch of biscuits. Two loads of sumac will work in a pinch, and the alder you've weeded out in the spring and dried under cover is good come winter. Cottonwood is good too, but needs a long drying period.

The oven can overheat easily and will with regularity till you get to know it well enough to cut down on fuel or bottom draft at the right time—well in advance of the time on the clock when you desire the heat change to take place. To cool it, restrict oxygen flow using dampers and draft ports as with any stove. Opening the oven for a few seconds works well with some dishes, while others such as cakes or bread will fall flat. So, keep a few bricks, flatirons, or a couple of big stones handy. Put quickly into the range, they will absorb heat and cool the oven. How many of what to use and how long to keep them in is another trick that each cook will learn only from experience.

The brick or flatiron trick is to achieve a quick heat reduction. For a normal two-temperature dish such as popovers or Yorkshire pudding, you can reduce the starting high fire to moderate by opening the draft at the top of the firebox to let more room air into the stove, close the lower draft and wait till the fire dies down to the heat you want for the second moderate cooking temperature.

Here's how Louise cooks the Yorkshire pudding while I'm tending the roast beef on the roasting spit out in front of the open-fire stove.

### Louise's Wood-Cooked Yorkshire Pudding

When the roast begins its two hours or so of slow-turning in front of the open stove, the cookstove is put on "bake" and filled with a load of good coals-producing wood—hickory or beech or oak chunks.

In the big, copper, egg-beating bowl (the tiny trace of copper that gets into the eggs helps the whites puff up), Louise mixes a scant cup of unbleached flour and a half teaspoon of salt. She then adds a half cup of rich milk or half-and-half and stirs till well mixed. Then in go two eggs, one at a time, the batter being whipped till frothy after each with the big wire whisk. A half cup of water is added, and the batter is whipped till bubbly, covered with a damp cloth and put aside.

At the end of the first hour of roasting, the pot beneath the meat will contain a good supply of fat and drippings. This is put on the cook top for 15 minutes or so to heat till sizzling. The firebox is filled with biscuit wood and both the bottom draft and the damper are opened full. The oven will reach Hot, or about 400 degrees F. by the time the fat has become smoking hot. The batter is poured in and the pot is popped into the middle of the

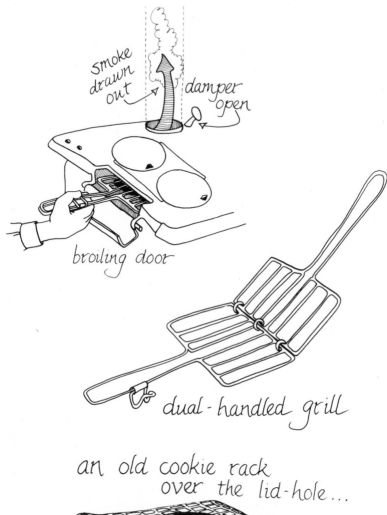

smoke drawn out

damper open

broiling door

dual-handled grill

an old cookie rack
over the lid-hole...

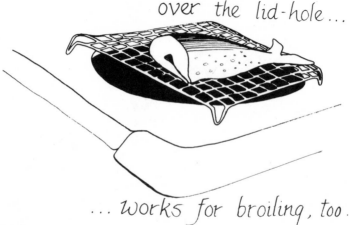

... works for broiling, too.

oven. The hot fire is kept going, with more biscuit wood added if needed for 15 minutes. The pot is given a half-turn every five minutes during this time so it will heat evenly. (Breads of all kinds need a quarter-hourly turn, in our oven at least.)

At the end of the hot bake, the firebox is filled with quarter-splits, pieces of a six- to eight-inch-diameter log that has been quartered. Now the fire is "checked" to reduce the temperature. This can be accomplished in several ways on our stove. The front doors of the firebox shaker and cleanout may be opened. So may the top broiling door or the air slots in it. The bottom draft control can be closed, and the damper in the boot may be half closed to the "check" setting or closed entirely. Just which measures are used depends on the strength of the draft that day, the wood and all those other imponderables. Usually Louise checks by half-closing the damper, partly opening the upper draft control and closing the lower control halfway. The oven is kept at "plain cake" or about 350 degrees F. for another quarter hour or so, baking till the pudding is well risen and golden brown on top. Then we are treated to beef and Yorkshire pudding, with a green salad and hearty red wine, a classic meal harking back to old England's open hearth cookery.

### Accessory Uses of the Oven

Broiling with the range is the same as in the open fire, except that you need a long handle on your grill. They still sell dual-handled grills for use over campfires. Build up a strong bed of coals, open the draft and the damper full, and remove the top lids and center divider. The meat can be seared through the broiling or loading door if you have one, or just poked in through an open top lid. Some old-time cookbooks recommend actually laying the meat directly on the coals, though I can't see how you could avoid getting ashes in the main course.

Once you've had a warming shelf or ovens above a wood-burning range, you'll never imagine how you cooked without them. Plates warm perfectly there, and many cooks just keep the dinner plates in the warming oven when they aren't being used. Bread rises perfectly, salt stays dry and loose, yeast bubbles up quickly, and any dish you want warm but not hot, such as bread or leftover apple pie, warms up in jig time.

Don't neglect to use the towel racks—one or two swing out

*family center...*

from the left side of most ranges. If your old-timer is missing them, you can fashion new ones to fit in the holes in the left-hand rear of the stove top. They are meant to hold the dish towels for drying, but they are fine for snow suits, mittens, and the weekly laundry, bit by bit.

Once again, let us apologize for being so vague about the specifics of using the oven. But no two bakes are completely identical. Louise has repeated a huge success detail for detail on a following day and had disappointing results. So if she tried to give you advice on whether to bake your bread on the bottom of the oven, or on the rack put on first, second or top setting, it would be downright pointless. You simply have to try each in its time. But once you master the range, you can be proud of a genuine skill. And, proud in the knowledge that *you* baked your bread, cake or pie. It was all your doing to wrestle that cantankerous machine into putting out a perfect job, without one whit of help from the technology of an automated zombie of a range.

towel racks
slip in...

...when not in use

## Slow Cooking

Probably the greatest glory of the wood range is its ability to top cook more slowly—more gently, really—than any other device known. While the soapstone griddle is heating over the firebox, there's a cast-iron skillet set to one side with a rasher of thick-sliced bacon strips gently rendering. By the time the griddle cakes are done, we have bacon that is cooked through, crisp and browned on the outside, but still full-sized, or almost. Not all fried to a crisp with all fat and moisture driven out. Same's true of many other dishes—toast, for example. We just set a row of home-baked breads sliced thick on the hot-but-not-real-hot section of the cook top for two to five minutes per side. The gentle heat toasts the outsides to a golden brown so it is wonderfully crisp, while the inside remains soft and moist. Nothing resembling toaster-done toast that's either raw or dried out. Louise puts slabs of butter on the first-toasted side, and lets it melt in while the other side toasts. Friend, you haven't tasted toast till you've had it straight from the top of a woodburning range. (Though any stove will do a passable job, only a big range offers the wide variety of heats for complete flexibility in slow top cooking.)

Slow cooking of stews and the like is enjoying a new popularity—powered by electricity. You've seen ads for crockpots, insulated cannisters containing a low-power electric heating coil. You fill the pot with stew parts, turn on a timer in the morning and that evening it is done. But that isn't genuine slow cooking as you can do it on a wood range. The back of the cook top is hot eight to nine months a year—boiling hot during a biscuit bake, but usually just warm enough to keep a pot or tea kettle barely bubbling. Not hot enough to make the bottom of a stew stick or boil the fluid off, just a gentle slow simmer. There's nothing like it from any modern appliance.

The stockpot sits at the back of the range, the same brew for a week at a time—maybe longer. Into it go all the meat scraps but smoked products and fish. From time to time melted fat is ladled off to be saved for soap. And whenever Louise needs a quick soup, she adds cut vegetables and barley to several cups of

the
Stock pot...

simmers
all day
and thrives on
what
might have been
wasted

stock and boils till done. The stock is also the base for gravies and stews.

Slow cooking—simmering a bit faster than the stockpot, perhaps so that a pair of bubbles pop up every second or so—is *the* way to cook up an old rooster or a chunk of that grass-fed beef that didn't turn out quite as tender as we'd hoped. And words can't describe what the low, slow and mellow heat of a medium-fired cook top can do to a brace of pork or lamb chops, lightly breaded by dipping in whipped egg, then in seasoned bread or cracker crumbs. Cut enough fat from the rinds of the chop to render into a thin layer of melted fat. Put the chops on, jiggle them back and forth a few times to assure they don't stick, then let them sit there for 30 minutes to an hour per side—just barely popping. Turn when the first side is golden brown and repeat. You'll never fast cook a chop again. These are moist and tender throughout, cooked through if you want (you must with pork), or with lamb or veal chops, pink in the middle. But all are unmatchable. You can have the same effect with any grilled food—potatoes, breaded zucchini, chicken or fish. Slow frying or grilling over gentle wood heat is a kind way to cook food, and the food responds in kind. Just now, Louise is putting the final touches on a pot of slow-cooked ham and beans that has been back and forth to the range since noon, day before yesterday. Here's her recipe, and I guarantee no digestive problems: you can swallow the beans whole.

### Louise's Slow-Cooked Ham and Beans

When the hambone is showing through along most of its length, take a pound of dried beans, sort and wash well. (We grow our own beans in with the sweet corn; last year the variety was Missouri Wonder, a fat, brown mottled bean from the Midwest.) Put the beans and three quarts of water in a sheet metal pot. (Cast iron isn't good for soaking beans; it tempts rust.) Add a whole bay leaf, as many whole peppercorns as you like and salt—a pile the size of a quarter in the palm of your hand. Put the pot on the hottest part of the cook top, bring to a hard boil, then cover and push back to simmer for a quarter of an hour. This is to start the softening and to sterilize everything so the beans won't ferment and turn into bean beer.

Set the pot in a chilly place to soak overnight at least. A day and night is better. Come morning of the day you want ham and beans for supper, bring the pot to a boil with the morning fire and keep it just barely simmering till midday, for five to six hours. Then cut up in bean-sized chunks and add: one big onion, two carrots, both peeled, and a stick of celery, top and all. Top up the water, taste and add salt if needed. If you like a tomato-y soup, put in a quart or two of tomatoes, home-preserved in a thick tomato broth, or a big can of tomato paste—more if it suits. If not all like the tomatoes (some of us do and some don't), those who do may add in as much of your good, home-done ketchup as they like at supper. Oh yes, put in the hambone, meat on, and most of the fat cut off.

Now bring the pot to a soft boil again, then put it on the back of the stove and keep it so that a steady but very gentle flow of bubbles comes up. You'll have to check frequently if baking to make sure it doesn't get so hot as to boil and stick. In another five to six hours of slow cooking, the beans will be soft, tender and perfectly delicious. Before serving, boil the broth down as far as you want. Louise keeps it a fairly thin soup; the leftovers are grand for lunch.

### Maw Wilson's Corn Cakes

With bean soup you simply must serve corn cakes and buttermilk. You can find your own buttermilk, but here is how to make corn cakes—sort of a cross between corn bread, hush puppies and griddle cakes—a concoction out of south Missouri via Louise's mother, good Ol' Maw Wilson. For four moderate eaters, with a few cakes left over to toast for breakfast:

Heat the soapstone griddle to low griddle cake temperature, so water just skitters off. Then combine about a cup of stone-ground yellow cornmeal with a quarter cup of unbleached white flour and a good shake of salt. (You can vary the cornmeal/white flour proportions to suit yourself. Maw Wilson relies on pure corn.) Add in an egg—two if they are small, two tablespoons or more melted butter, ham fat from the bean pot or any other liquid oil and a heaping teaspoon of baking *powder*, not the soda. Now stir in enough whole milk to make a griddle-cake-consistency batter; a spoonful poured on the griddle doesn't sit in a lump, nor does it run all over but spreads out into a nice pancake shape. They cook just like pancakes. These proportions aren't exact since no one in Louise's family has ever measured ingredients exactly, and when I ask what they mean by a "good shake," say of baking powder, the reply is: "Well, a good shake, but not a *real* good one, you understand. See?" And they hold out a hand with a mound of baking powder in it. "Just about this much." And that's cooking with wood.

By the way, that meal of corn and beans and enough meat to flavor it, perhaps with a bit of greens on the side, provides perfectly balanced nutrition—the corn/bean combination of complementary proteins is better than either ingredient alone, and you could live on it. Indeed, it came to south Missouri with the first settlers who felt themselves lucky indeed to be able to live on it. Of course, the European settlers got the goods from the Indians; all the corn and many types of beans are native to the Americas. Corn and beans are still the staple foods of a lot of folks from our own Southwest, clear into Central and South America. So, savor a bit of real all-American history and cook up a meal of beans and corn cakes on the wood stove. And it may not be all history. Some experts are predicting times as hard as pioneer days ahead as the world runs short of everything you can think of except hungry people. But those of us with a few acres cleared, and a good store of corn and bean seed, our iron pots, a stove and the wood to cook with will sail through. A bit slimmer than now perhaps, but well fed and warm. Good cookin!

## The Range as Space Heater

It was a cold February when we moved in off the farm to the town place with the old Glenwood in the kitchen. Our second week the central heating went on the blink over a long holiday. All we had to keep us from freezing was the range, and she did. The upstairs went a bit chilly around the edges toward dawn, but that's what blankets are for. So when you read some expert telling you that cast-iron ranges are poor heaters because the fuel box is tiny and only takes twelve to sixteen-inch logs, you can bet he's never used one. The range is a tremendous heater, simply because of that great mass of cast iron soaking up and radiating the heat. Agreed, the fuel box is small compared with a big airtight heating stove. But that is so you can regulate your

heat; the small fire cools or heats in a matter of minutes—
essential flexibility for the cook. But fed often enough, it heats—
the Glenwood kept us warm all by herself for three cold days.

From the morning breakfast fire till the evening dish towels
are hung on the drying rods beside the firebox, the range has
been heating the kitchen and adjacent rooms, though her main
job was to cook. Now it's time to rig the fire to throw heat during
the night. There is always a good bed of coals in the firebox; I stir
these well, and if the ash level is high, I may take a few cranks
with the shaker to remove dead ash. I crank till a good "star
shower" of small coals falls into the ash drawer.

Then the firebox is filled, with the biggest unsplit logs I can
get in if the coals are hot, with quarter-splits on the bottom and
round logs on top if coals are low. We open damper and draft till
a good fire is caught, then all is closed down for the night.
Usually one of us is up once or more during the night with one
or another child, or a sudden urgent need for another cup of the
bean soup that stays hot beside the stockpot at the back of the
stove. The box is refilled if needed. Even without a midnight
loading the great hunk of cast iron retains heat. The kitchen

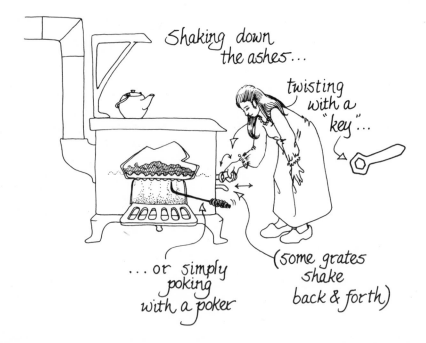

Shaking down the ashes...

twisting with a "key"...

...or simply poking with a poker

(some grates shake back & forth)

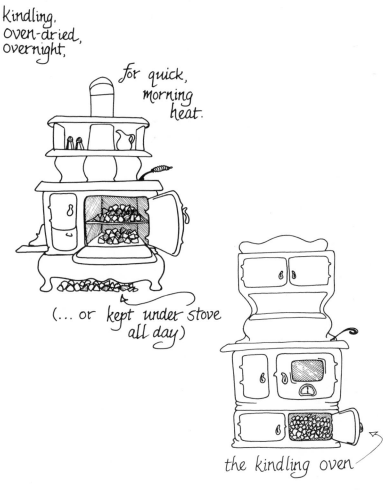

kindling.
oven-dried,
overnight,

for quick,
morning
heat.

(... or kept under stove
all day)

the kindling oven

seldom has a chill in the morning and is comfortably warm when
we come down except on very cold, windy nights. Even then
there is enough of a coal bed to get the fire going again for the
morning coffee and the teapot is still hot enough to brew up a
first cup. This is the best time to empty the ashes if needed. We
shake down till "stars" fall into the drawer. A handful of splinters
goes on, followed by a layer of small splits, not quite touching,
and three one-quarter splits on top: Louise's magic arrangement
once again. Draft and damper are opened and in ten minutes the
coffee water is boiling.

If the house is unusually cold we'll throw the oven control to
let hot air circulate through the whole range for a half hour or so.

in the morning...

... add splits to the still-glowing embers

This requires more draft, more wood for a hotter fire with the result that a great deal more heat is radiated into the air. Our oven door is always left open at night so the fire side of the oven will contribute the heat it absorbs. But the firebox alone is a good enough night-long heater—and on our range much more economical than when the oven is heated also.

So in addition to cooking your meals, the range will heat the lower story well and will keep the whole house comfortable in an

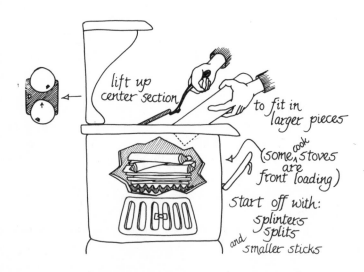

lift up center section

to fit in larger pieces

(some cook stoves are front loading)

start off with:
splinters
splits
and smaller sticks

emergency. We figure that we use about one-tenth of a cord (five dollars a week at most) in the range during a cold week for all cooking and perhaps a third of our heat. For heat alone, and using quarter-splits, we average one stick an hour for mild heat, one per half-hour for medium heat and two per half-hour to toast ourselves after an afternoon of tobogganing down the back hill.

# CHAPTER SIX
# Getting the Wood In

Now that your fireplace or stove is in and operable, you'll need some wood. It's easiest to buy it from a woodlot operator who has the equipment to turn out cordwood on a commercially profitable scale. Look for ads in the newspapers of the nearest rural towns. This is not to say that the fellow who puts up a cord or two on weekends should be avoided; likely he will give you a better price than the full-time woodman, but you'll have to ask around to find him. Besides, a man who has the tractors and logsplitter and all has a considerable stake in getting and keeping customers. By and large, if he tells you his wood is good, it will be, and he won't try to pass off newly cut wood for well aged, or overaged and punky wood for good wood. He'll probably be particularly friendly if you let him know you will be in the market for a half-dozen or more cords a season. But you should still know what you are (or should be) buying.

## Selecting the Best-Heating Woods

First, if at all possible, insist on hardwood if you can. Soft coniferous woods such as pine are lightweight and resinous. They will burn; in fact, they make the best kindling there is. But most species last no time at all and can get up a roaring blaze that not only wastes wood but can very quickly overheat a stove. Too, the resinous sap will not burn completely, unless the fire is too hot for best economy and safety. The unburnt sap will deposit on your stovepipe and flue as creosote. Old-time New Englanders never burned pine and similar woods out of justifiable fear of flue fires.

Of course, the species of wood you get depends on where you live. In much of the South, you'll be burning pine because that is about all you have available. Same is true in the co-

300

niferous forests of the West, but in the most heavily populated areas of the cold belt, hardwoods are available. In the chart "Characteristics of Several Wood Varieties" we've listed comparative weight, splitability and such for the most common woods in the East and Midwest. In our part of New England, rock maple, birches and hickory predominate. Farther north, the woods turn to conifers; down South there are more beech and oaks, and the hickory gives way to black walnut. All burn well. One tip if you are new to wood and wood heat: look for firewood in the highlands. Bottomland trees run to cottonwood, sycamore and soft (red) maple that are watery and take a long time to dry. Some aren't worth much as fuel even then. Don't try to find any chestnut, though most of the wood heat books around dutifully list its characteristics along with other fuelwood data gleaned from out-of-date forestry publications. There hasn't been any American chestnut around since the chestnut blight cleaned it out several generations back.

## Buying Firewood

As we must have mentioned earlier, wood is sold by the standard cord, a pile of four-foot-long logs piled four feet high in a stack eight feet long. When you buy it, big logs should be split and cut to whatever length you need. Four-by-four-by-eight feet is a pile 128 cubic feet in volume no matter how you cut it, though. Dried wood weighs up to 63 pounds per cubic foot (for shagbark hickory, the most dense of common hardwoods). About a third of the volume of any cord is airspace and most woods are lighter than hickory, so your typical cord of well-dried hardwood should run around two tons in weight. Obviously, if someone tries to tell you his little half-ton pickup contains a cord of wood, he's gotta be kidding. Don't let him kid you.

Most professional wood sellers will have some sort of dump body on their delivery truck. I've seen one- and two-cord loads delivered by a big five-ton dump truck and half-cords delivered on a beefed-up light truck with a makeshift dump bed on the back. Some years back, Louise and I bought a cord of wood as stacked and transported it home in the back of a Volkswagen. It took quite a few trips. I'd advise you to have your wood delivered, even if it does cost an extra $10 a cord (as seems to be the case hereabouts these days). Remember, the heating value

Sam, Martha and their pal Christie are stirring up an honest half-cord of assorted hardwood limb pieces, cut to 12-inch stove length. Note the deep tire tracks on the lower left. This is early spring, mud time in New England, and the truck got bogged down and couldn't get back to the wood porch. The kids hauled the bulk of it to the porch stick by stick over the next few days and got a kick out of really helping. They keep the kitchen wood box full of kindling, too. Well, usually they do.

of a cord of well-aged hardwood is the equivalent of up to two hundred gallons of No. 2 fuel oil, about $100 at today's rates. If you spend $50 a cord for the wood plus another $10 for delivery—even another $5 or so to have it stacked, you are ahead of the game.

Actually, don't have it stacked or split into small sticks. Do the stacking and splitting and hauling yourself. It is easy work; no one stick weighs much. But piling up several tons of wood one or two sticks at a time is perfect exercise, particularly if, like most Americans, your regular work is sedentary. Splitting up and hauling in a hundred pounds of wood every evening gets your heart pumping and blood moving at a moderate pace for five or ten minutes every day. Then, there's the bending and shovelling and hauling of the ashes, most likely into the cellar several times a week. Honestly, keeping a couple of wood stoves fed and cleaned will supply about all the exercise you need to keep reasonably fit during the winter. Spring cleaning entails a lot of elbow grease too. Any wood-heated house picks up a covering of the gentle rain of ash that escapes every stove or fireplace I've ever seen at each loading and clean out. Not "dirty dirt" as my Grandmother used to say, but still requiring a good scrubbing now and again. No doubt about it, wood heat is hard work from start to finish. But if you take it up, then garden summers like the fellow whose old stove we refurbished a while back, you may live as long as he did. He died prematurely at 99.

### Reconnoitering

If you've reason to suspect the reliability of your wood seller, it's best to visit his yard before you buy. Chances are that he won't have wood sitting around in neat cords but stacked here and there in piles. An experienced wood stacker can tell when he has a cord or half-cord on his truck by the look of the pile or the settling of the springs. Besides, your cord and mine will never contain the same amount of wood. Back when charcoal burners bought cords of wood where and as stacked, the loggers worked out some pretty dandy ways of building the maximum amount of air into a cord. Take the seller's word, restack back home if you like and complain if you feel you've been short-sticked. Chances are you'll have the law on your side before long. Several states are already trying to define what a true cord is and will be regulating wood sales before too long.

### Drying Firewood

Do check the wood for dryness, though. Even fully air-dry wood will contain 20 to 25 percent moisture. However, some green woods are over one-third water by weight. You want as little of this moisture as possible in your fire. Green wood is hard to light. Too much heat energy is lost turning the moisture into steam and driving it out the flue, where it contributes to creosote formation. Check the cut ends of a sampling of the wood. Green wood will show the growth rings and perhaps a surface roughness from the saw, but the cut face will be solid. As wood dries, it shrinks in circumference—never lengthwise. Shrinking takes place unevenly and stresses occur in the wood, resulting in small cracks. Most species show cracks running from

Here is a stack of well-dried cord-wood. Note the cracks that indicate dryness (the wood is six months off the stump) are different sizes and come in different patterns depending on tree species and age.

And here, for comparison, are two sticks off the same small (quince) tree. The year-old branch is dark-colored and well-split in the characteristic ray pattern. The day-old stick shows the beginning of dry-cracks in the center. The cracks will expand out ever so slowly as the wood dries.

the center to the bark. Others will crack along the annual ring. Still others will show random checking. With a bit of experience you will also be able to distinguish the dull thud of one wet stick being knocked against another from the slightly crisper sound of dry wood.

The air-drying process is a function of time, taking place faster in warm dry seasons than wet or cold. Basically, the wood has to lose its inner moisture to the less saturated air. Split wood dries half again as fast as whole logs, some species faster than others. If the wood has been cut and split, then aged over summer it can be as young as four months and burn well. If piled in four-foot unsplit lengths in the woods, then cut and split, it should be the better part of a year old.

Sitting out in rain or snow has little effect on the wood's essential dryness, except for the top and ground layers, which may have wet bark. Any surface moisture will evaporate overnight so it's a good idea to get all of tomorrow's wood in this evening. Stack kindling in the kitchen oven or under the range.

## CHARACTERISTICS OF THE BEST COMMON FIREWOODS

| Hardwoods | Efficiency: wt./1 cu. ft., dry | Splits | Amount of Smoke | Wet, Medium, Dry |
|---|---|---|---|---|
| Hickory | 63 lbs. | v. well | little | m |
| Black birch | 49 | fair | medium | w |
| Pecan | 46 | well | little | d |
| Beech | 45 | hard | little | d |
| Sugar maple | 44 | fair | not much | m |
| Black oak | 43 | fair | little | w |
| White ash | 42 | well | little | d |
| Yellow birch | 40 | hard | a bit | m |
| Red maple | 38 | fair | some | w |
| White birch | 37 | easy | some | w |
| Black cherry | 36 | fair | little | d |
| Tupelo or gum | 35 | poor | not much | w |
| Sycamore | 35 | hard | some | w |
| American elm | 34 | doesn't | plenty | w |
| **Softwoods** | | | | |
| Southern yellow pine | 40+ | v. well | a lot | d |
| Eastern red cedar | 32 | nicely | some | d |
| Cypress | 28 | so-so | medium | w |
| Redwood | 25 | fair | medium | d |
| White pines, firs | 24+ | easy | some | d |

## Mixed Cords

Now, you shouldn't expect a cordwood seller to bring cords of only the best-burning species. His woodlot will contain a mixture of species, and he'll cut and sell them all. If his cords appear to consist mainly of the low-weight kinds such as white birch and black cherry, offer him one-quarter less than he asks, particularly if you know another dealer who has a mixture with more maple, oak, or other dense wood.

Especially beware of a seller offering you cords of American elm. The wood burns alright, don't get me wrong. But it is lightweight when dry—contains over one-third moisture by weight when green. Also, it is almost impossible to split. Not only is the grain all twisted around, it is highly elastic and springy wood. Every old-timer out our way has his story of the greenhorn sent out to split an elm log who was last seen by human eyes, whacking away at the log as it bounced along into the woods ahead of him. Further, elm may be acting as a nursery to bark beetles, a million or so of them to each recently dead tree. If you store the wood inside, you may one day find yourself godparent to a whole lot of harmless, but bothersome

a woodshed

little bugs. Actually, if elm has been aged for a proper length of time, any bugs will be long gone. It's the same for other species. The chances of importing a nest of termites or carpenter ants, for example, along with the wood pile is nil. You'd have to be unlucky enough to get the brood chamber and fertile queen, which in most species are well underground, not in the wood at all. The only pests we've ever had in the woodpile were a nest of bumblebees and innumerable field mouse nests, but they all moved in well after the wood was stacked. I wouldn't worry about bringing any harmful critters from the woods into your woodshed.

## Storing Firewood

Speaking of a woodshed, if you have one, great. In our various homes we've stacked wood in a genuine woodshed, a slatted corncrib, the cellar, on and under a big porch. Store your wood where you have the space. The major consideration is to minimize time and distance in hauling. Ideally, you'll be able to cut wood into stove or fireplace lengths in the woods, toss them directly into a truck, then restack in a shed just a few steps from

a conveniently placed woodbox

the back door to dry under cover for a year. That requires a lot of stacking room that we don't have anymore.

We cut wood into standard four-foot cord lengths and leave it in ricks in the woods over winter, then cut it to size and haul it in several times over the fall and winter. The back porch is just outside the kitchen and bringing in the wood is a short walk (that the kids enjoy most evenings). I'd advise you to work out the easiest, least time-consuming system of bringing in the wood that you can before you take on the first cord. If you simply have to store it at a distance, get a big garden cart (the biggest and best I know of comes from Garden Way Research, Charlotte, VT 05445) and pull several days' requirements up to a convenient door or window every so often. Having to hand-carry armfuls over long distances several times a day will add too much labor and, in time, boredom, to your wood heat experience.

Louise brings in the first sticks from a newly stacked cord of mixed hardwoods. There are three faces of foot-long logs at the rear, a tenth of a cord of quarter splits (a week's supply for the range) to the left, and, in the box, several day's requirement of small stuff for kindling and biscuit wood.

### Arranging the Woodpile

It may seem like a minor detail, but when you are stacking your wood, it's a good idea to divide it into "grades," particularly if you have an unlighted and unheated woodshed and, come winter, will want to get out and back with the type wood you need quickly. I try to separate wood into four categories, either in separate piles or separate sections of the pile.

Number One is the quick-burning stuff in order: splinters, small kindling splits of anything and quarter-splits of light woods such as the birches. Number Two is the same woods, but in half-splits or small logs. Number Three is composed of whole round logs of the light woods plus quarter- or half-splits of harder woods. The fourth, and really most important from a wood-garnering-at-midnight point of view are whole logs (any size, but the bigger the better) of the really dense, long-burning woods. When the fire gets low at 4:00 A.M. of a −30 degree F. winter evening, you want to get out to the woodshed and back quick, and with the most firepower you can grab. So in pile Number Four go the hickory, hornbeam, oak and other such logs that will burn for a good four hours. Stack it so you can get it blindfolded and when you have to, you can.

# Tips for City Folks

If you live well into a metropolitan complex, away from the woodlots, you'll likely run into one or another of the tricks some of your crafty country cousins like to pull with their wood. One is to charge you double or triple the price a cord would bring a half hour drive out of town. The winter of 1974/75, first full year of the "energy crunch," wood was bringing $100 a cord in Boston. Maybe worth it if you have more money than time to spend and only need a few logs for their entertainment value. But if you are serious about heating with wood, it will pay you to get out into the woodlands and scout around for good wood at a fair price. I'd say, spend one weekend scouting, and if you can't find a wood dealer who will deliver for a reasonable fee, then the next week rent a really big truck (one of those moving vans would do, though a dump truck is best) and haul the wood in yourself.

Be sure to get well out of town. Louise and I did a wood survey in the Boston area not long ago. Right in town there were trucks loaded with puny little elm and white birch limbs, being sold by the half-bushel basket full: two dollars for the elm, three for the birch. The sellers were hawking "cords" of wood. A little way out of town, in the high-income Concord/Lexington area a "cord" consisted of a wooden frame about two-feet square filled with fair-sized limbs. Five dollars for that. Even farther out, but still in the suburbs, a fellow was selling what I'd call a "face cord," though it was an honest cord to him: a stack four-feet high and eight-feet long, but only a foot deep. A quarter-cord, in fact. (A face cord is any four-by-eight foot stack of logs cut to burning length—could be any depth.) This man wanted $25 for his quarter-cord, and the buyer had to haul it off.

You may just run into someone who insists that his cord, three feet by five feet by six feet is a legal cord. He's legally correct; 90 cubic feet of wood is a cord according to people who measure heating values. It must mean 90 cubic feet of solid wood, though. The difference between that and 128 cubic feet in a conventional fuelwood cord represents the air. If anyone tries to sell you one of these legal cords at regular cord price, $50 and up this year, offer him 70 percent.

## Cutting your Own

There are plenty of places you can look to get your own wood, free for the hauling, or perhaps for a small fee. More and more federal and state parks are being opened for supervised cutting. I'd say, scout around for such opportunities, perhaps by checking with the state forester (at the extension service of your state's land-grant college). There's at least one in every state. Vermont has a forester in each county, or so I'm told.

Any heavily used government-owned wooded areas, including roadside picnic areas and parks in town will be frequently culled of dead trees and limbs, if only to remove a potential hazard to visitors. Check with the street and park department. If they won't let you cut down trees, you may be able to pick up windfalls in the spring. Any area where American elms grow (grew) will have an elm removal unit. Most are professional landscapers contracted by the city. They come in with heavy equipment and take out a big tree in half a day. You may not be able to pitch in, but you can find out where they take the elms and any other trees removed. Most places, the elms are supposed to be buried or burned, and you should be able to get permission to take out as much of the smaller wood as you can.

## Sources

An excellent source for one of the best-burning woods there is is any nearby fruit orchard. Most fruit trees are pruned annually, and a good orchard keeper is constantly culling diseased or too-old trees and planting new stock. Most will let you take out the old wood and if you do a good neat job, they may even pay you for doing it.

Nurseries, landscapers and the local "tree expert" are all possible wood sources. In logging country (and you'd be surprised how much small-scale logging activity exists near population centers) you can find good fuelwood at the sawmills or out in the woods. Loggers or landowners are usually only too happy to let you clean up the tops of trees cut for lumber. Sawmills will have huge sawdust piles and if it is a mill that processes logs fresh from the woods, there will be great stacks of slabs, the bark and outer wood removed to square up the logs. In our area, the slabs are strapped up in approximately cord-sized bales and sell for around $15 each. You get a lot of bark and air, but it burns and it is cheap. Be sure to get hardwood slabs if possible.

Any factory or mill that works with wood will have scrap. Foolishly, in my opinion, clean air ordinances may keep them from burning it themselves for heat, and you may be able to get chunks of beautifully aged cabinet or other milled wood, possibly free for the hauling.

## Buying a Woodlot

Finally, you may want to buy standing timber or a woodlot. Landowners often sell timber on the stump for $100 or $200 an acre; this would be for all the timber of marketable size. The owner will expect you to leave the small stuff to grow up and net him another few dollars in a few more years. Perhaps you can get permission to cull out only dead or diseased trees for the wood, and if you pay for the privilege—or if you pay very much—you are getting the worst end of the bargain. Purchasing the land itself purely for the wood may make sense in a lot of places where unimproved land is cheap and plentiful. Make sure that land borders a public road or you have deeded vehicular access, a right-of-way in. A plot of 12 acres or more, well treed with a variety of hardwoods, should turn out all the firewood you can use at an average rate of one cord per acre per year—and it will keep on producing indefinitely. Be sure to make an informal timber cruise of the land before you spend any money. Go over it all in parallel paths, so you see every tree on it. Of course, you want a majority of dense, good-burning hardwoods. If there is a lot of big timber, you'll have a hard time getting it cut into fire-wood. It's best to find a stand of young trees (as we do in cutting our own), and plan to leave the big logs or have them taken out by professional loggers with the necessary equipment.

In cutting your own, you'll need a chain saw and some other gear mentioned in the next section. I'd say, rent it if you are only spending a few days a year in the woods, getting in a cord or two. Good saws cost $250 to $500 and they need a lot of

Sections for cutting stove-length sticks

Sawbuck

(What little boy couldn't spend hours in the fragrant sawdust under his father's sawbuck?)

## Bow-Saw
### (Large)

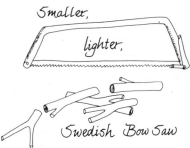

Smaller,

lighter,

Swedish Bow Saw

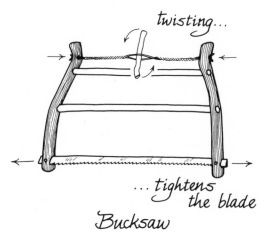

twisting...

... tightens
the blade

## Bucksaw

maintenance. It's best to let the rental agent do it unless you are really into wood on a significant scale. A couple of good men with rented saws and a borrowed or rented pickup can get a whole lot of wood out of a park or an orchard in two or three weekends.

In most any kind of gleaning and in working your own woodlot, there will be a lot of little branches you shouldn't waste. The small stuff is usually flexible and willowy enough that a chain saw just wiggles it around, without cutting it. Probably the easiest thing to do is cut it into handy four-foot or so lengths and store it that way till needed. Then to cut it up, build yourself a sawbuck, traditionally a pair of X-shaped ends attached to each other with two or three longitudinal stringers about a yard apart. Ours, for cutting 12-inch stovewood has three Xs, one just under one foot from one end. That way I can cut any length log we need. For sawing up the little stuff, spend a few dollars for a 30-inch Swedish bow saw. Blades are ultrathin and sharp and cheap to replace. There is an antique bucksaw and a two-man crosscut out on the farm, but both take wide cuts and are heavy as can be. When woodcutting time comes around, it's not so bad living in the mechanized 1970s after all.

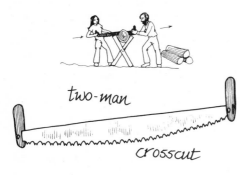

two-man

crosscut

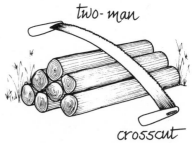

two-man

crosscut

# Managing the Woodlot Naturally

Most books and government publications on woodlot management that I've seen tell you to grow the maximum number of marketable sawlogs. A mill will pay from $10 for a mature white birch to several thousand for an especially fine black walnut log. There's a whole profession, forestry, geared to thinning, pruning, and selective cutting for sawlog production. And the industry set up to use sawlogs and forest products rivals the oil companies in power to work up depletion allowances, special tax breaks and freight rates. (It's cheaper to ship new pulpwood for paper than to ship old paper for recycling.) Well, foresters are good men, outdoorsy and sincere in their work. And if I'm let rant on about the abuses of the industry you'll get bored and I'll likely get apoplexy.

### Retaining the Natural Mix

We and our woodcutting friends feel that the woods ought to be let grow as naturally as possible. Every tree has a purpose, whether we understand it or not. We intend to retain the natural mix of hardwoods, coniferous trees, scrub and meadows that nature left us. It just seems that in the long run, thinning a forest to the few species and tree shapes and sizes that make the most profit for the industry is as unnatural as growing nothing but corn on your farmland year after year. Monoculture is monoculture, and with trees as with food plants, it is asking for epidemic disease, soil imbalance and an upset ecology that will ultimately call for corrective measures such as the sprays they are using to control gypsy moths in the artificial in-town plantings of the East and the several recent insect epidemics in the single-species lumberwood plantings of Douglas fir and the like out West. Nope, we let nature mix up the trees as she will. I guess we're organic foresters.

Our woodlot has been logged three times, twice since the turn of the century, and still it bounces right back with enough to spare that we can cut our fuelwood supply and never make a dent. And that's the way we'll keep it—we and the trees and all else that lives there, existing in natural harmony, us letting nature make the big decisions and nature letting us remove enough wood to keep us warm. The oddly tall maple across the road will remain: Baltimore orioles nest in the highest branches every spring. The sunny, but scrub-choked glades in the woods will stay as they are: mice nibble the lower few inches, rabbits feed in the next foot and few inches, and deer browse the upper layer.

I know for a fact that rabbits use the humps and hollows of a huge "unthrifty" rogue white pine's roots for winter quarters. A whole succession of them some winters, as the live traps turn up a good rabbit stew from the tree once or twice a month in snow time. I'm almost as sure that the same quarters are home to litters of one or another critter each spring. That's what the dogs say, snuffling around the needles each time we pass the tree on a walk.

By the rules of modern forestry, this old tree would be one of the first to go. The term rogue applied to timber means a tree that sprouts lateral branches low down on the trunk, so the sawlog is too short. New England is just now growing back from the deforestation that commenced in colonial times and continued till the rocky hillsides became uneconomical to farm in the 1930s and 1940s. Even in the early days they considered "rogues" too poor to cut. But they didn't cull them. They left them grow to shelter wildlife and act as seed trees.

We find one of these out-sized, odd-shaped giants every hundred yards or so in the woods. The dogleg along our western boundary was marked by a huge chestnut till the blight struck. The stump's still there, sending up shoots that are blighted after a year or two. And there is a fantastic lightning-split black oak where the deer paths meet up at the north corner. The tallest American elm I have ever seen is down by the road. We'd hoped that it might be one of the few to hold out against the Dutch elm disease, but last year the leaves browned early up in the crown. Bark is sloughing off the upper limbs now, and Louise says the highway department has affixed one of its metal tags scheduling the tree for removal. Barring such disease, fire or storm, though, all of the big trees will remain. They are what make our woods a forest. It's their sons and daughters that we use for fuel.

### A Cord an Acre a Year

Now, the old-time rule of thumb is that an acre of land will produce about a cord of wood a year indefinitely. But don't plan

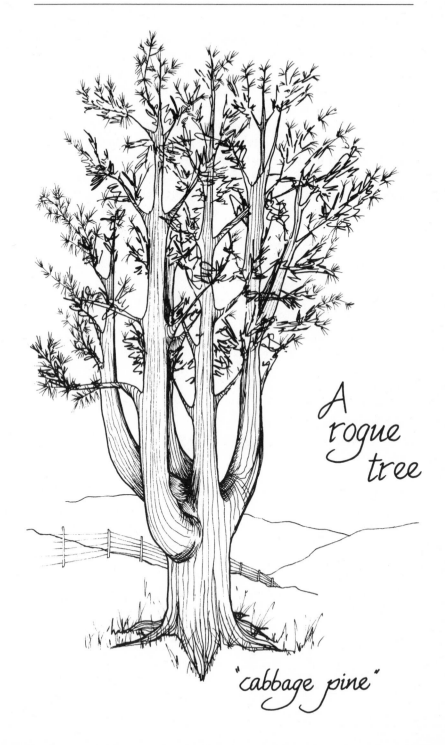

A rogue tree

"cabbage pine"

to bull your truck or tractor over your whole woodlot each year to extract one cord from each odd acre or so. Most of our own woods is left alone; fuelwood is cut from several scattered stands of the right-sized trees. For sheer ease of handling I prefer them "half-cord" sized, with bases of perhaps a foot to eighteen inches in diameter. The trunk of a tree that size will make up about a quarter-cord when cut and split, and the limbs and larger branches will fill out the rest of the half-cord. A few trees such as white birch seldom grow much over good cordwood size, whereas a maple, oak or ash is a mere baby when still that small. Eight to twelve trees of the proper size will supply our year's wood needs.

A ½ cord Tree

Of course, our initial cutting on the woodlot was to clean out injured and diseased trees and gather up any fallen limbs that were recent enough to be unrotted and burnable. No reason to waste good wood. We worked especially hard to remove diseased elms. In those days it was believed that the bark beetle that carries Dutch elm disease could fly only a short distance, and "tree-hopped" through the woods. Supposedly, if you cut down the infected trees, others nearby would be spared. Recent research shows that the little bugs can fly up to five miles looking for a healthy elm to feed on and a diseased one where they will lay eggs under the bark. So, from a prevention standpoint, culling sick elms may be a waste of time. However, the bleached, debarked skeleton of a long-dead elm is a saddening spectacle in the woods, and we are glad that ours are gone.

### The Modest Clear-Cut

In the process of getting to elms, split birches and the occasional maple showing the white, hard shelf-like fruiting members of a common decay fungus, we found we were tearing up the woods with the truck and felled trees. That's when we decided that taking that one cord per acre each year wasn't

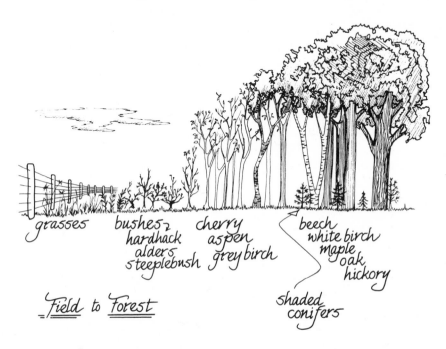

grasses    bushes    cherry    beech
           hardhack    aspen    white birch
           alders    grey birch    maple
           steeplebush           oak
                                 hickory

Field to Forest           shaded
                          conifers

practical. Now that the larger culls are gone, we use a form of clear-cutting for the firewood supply. We choose a good stand of proper-sized trees and go right through it until it is cut out.

In the old days the tillable parts of New England were denuded of all but rogues, a few pasture oaks left to shade the cattle, and trees along roads, fencelines and walls. Then, as fields were abandoned gradually over the last century, each was seeded by the nearest mature trees. Typically, there would first be a stand of quick-growing scrub or "nurse trees" such as aspen, alder, gray birch or sumac to shade the tender, slow-growing seedlings of the next generation, maples perhaps, or other longer-lived hardwoods. And as the hardwoods grow they in turn shelter seedlings of the shade-tolerant conifers that (in the colder areas of the continent) will replace them as they fall or are cut in the natural course of time.

So, we find many stands of like-sized trees of the same species or several that commonly grow together. Maple/-beech/birch is a common mixture in our area as is oak/maple/hickory. In the bottomlands we often find a few syca-mores mixed in with gray birch and more common hardwoods. Many abandoned farm sites still contain aged apple trees (finest firewood of all). Throughout the woods are scattered stands of American elm, and occasional white or black ash, ironwood (hornbeam) and black cherry trees, along with a rare basswood, persimmon, tupelo or sassafrass. When we find a white oak, it is marked for preservation. Too many of the species have been cut, and the wildlife so love the sweet white oak acorns that very few survive to propagate the tree. When and if we find an American chestnut or elm growing in apparent health, it is noted for per-servation and continued attention. Someday a mutation will come along to preserve each species from their respective diseases.

However, there are a dozen or so stands of the right-sized trees on the lot and each year I cut a road into a new section of a stand and cut out our five or more cords of wood—details on log-ging a bit later. But our objective is to cut out all trees of half-cord-size age and older. There will always be a profusion of sap-lings ready to replace the older trees, and once the big ones are down and stacked, we go through, cutting out runty or un-healthy looking saplings, any black cherries with black knot fungus, oaks infested with galls or young birches that have cracked under an ice load. We aren't too particular about which species we cut and which are left. Nature has already set up the most desirable (from her point of view) succession. All we do is eliminate the halt and the lame and cut down on unneeded competition. Normally in these stands, there will be seedling maples, say, already twenty feet high, but only a few inches through at the base. With both the big trees and a lot of the runts gone, they will have free sky and soil to put on weight. Then when we come back in 15 or 20 years, they will be nicely grown to cordwood and will come down—or about three-quarters of them will, along with all of the other species, if we've decided to enlarge the sugar bush. In the interim we'll have a sunny glade to picnic in. The scrub will grow up for the deer to nibble and the neatly piled slash—small branches and twigs left over—will shelter the rabbits when the snow comes. Thus we will repeatedly cut out selected patches here and there. Most of the woods will grow on undisturbed to become whatever type forest nature intends. If anything, the frequent little clearings and piles of drying wood offer a bit of variety to a walk in the woods.

4-foot logs
left to age

# The Woodburner's Guide to

The good, the not-so-good and the available are listed in this catalog of fuelwood trees. While it would be grand if everyone burning wood for heat had ready access to good-burning hickories, oaks and locusts, that simply is not the case. Elsewhere in this chapter is information on the heating qualities of various fuelwoods. The point of this catalog is to guide the

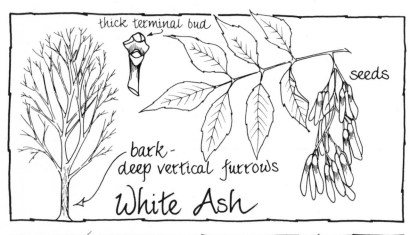

thick terminal bud

seeds

bark – deep vertical furrows

White Ash

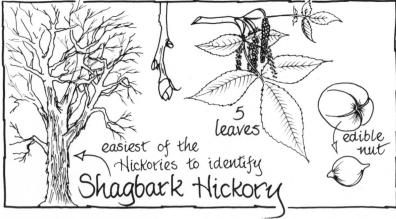

5 leaves

edible nut

easiest of the Hickories to identify

Shagbark Hickory

# North American Fuelwood Trees

neophyte woodchopper to the best and most common fuelwood trees. Remember that identifying the fuelwood tree is only a part of the task. It must be safely felled, bucked and split, properly seasoned and burned in a carefully maintained stove for the woodburner to fully realize the tree's heating capabilities.

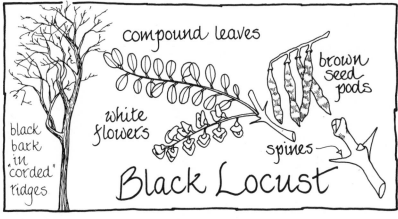

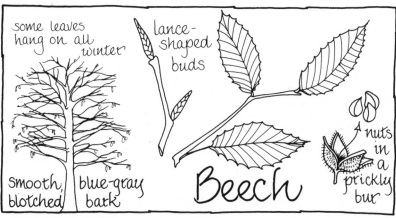

buds

too precious
for
firewood
if you
have a
sweet
tooth

deep gray bark
Sugar Maple

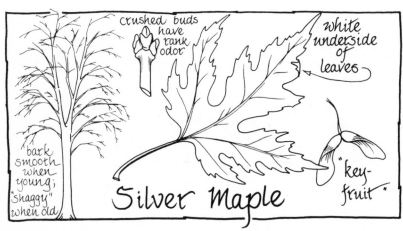

crushed buds
have
rank
odor

white
underside
of
leaves

bark
smooth
when
young;
"shaggy"
when old

Silver Maple

"key-
fruit"

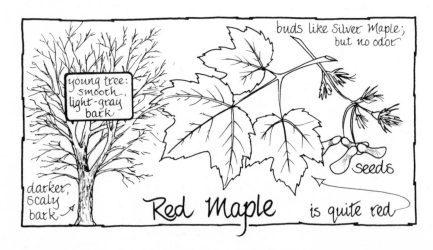

young tree:
smooth
light-gray
bark

buds like Silver Maple;
but no odor

seeds

darker,
scaly
bark

Red Maple

is quite red

bark—
shallow
ridges

Red Oak

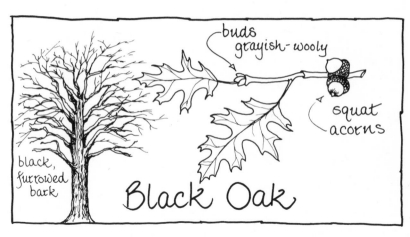

buds
grayish-wooly

squat
acorns

black,
furrowed
bark

Black Oak

scaly,
gray
bark

White Oak

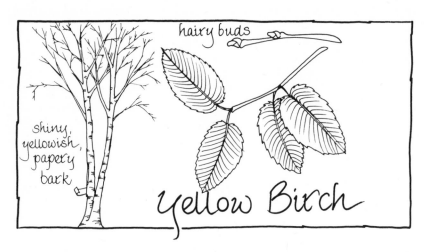

hairy buds

shiny, yellowish, papery bark

*Yellow Birch*

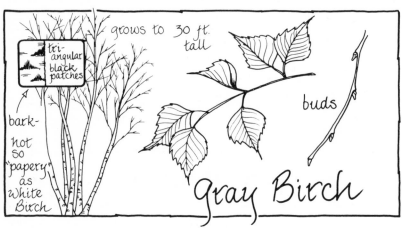

tri-angular black patches

grows to 30 ft. tall

buds

bark—not so "papery" as White Birch

*Gray Birch*

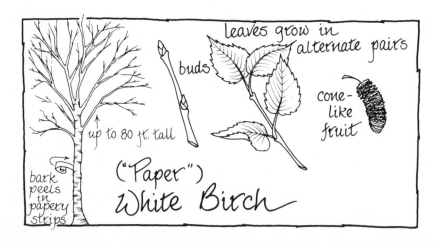

leaves grow in alternate pairs

buds

cone-like fruit

up to 80 ft. tall

bark peels in papery strips

("Paper") *White Birch*

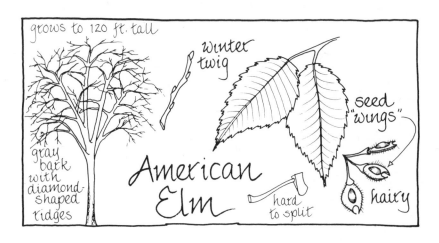

grows to 120 ft. tall

winter twig

seed "wings"

gray bark with diamond-shaped ridges

## American Elm

hard to split

hairy

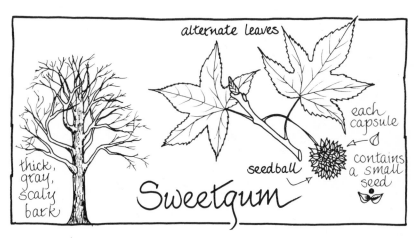

alternate leaves

each capsule

seedball

contains a small seed

thick, gray, scaly bark

## Sweetgum

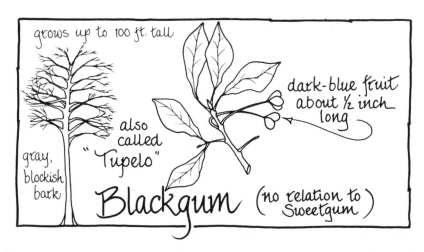

grows up to 100 ft. tall

dark-blue fruit about ½ inch long

also called "Tupelo"

gray, blockish bark

## Blackgum (no relation to Sweetgum)

a fragrant albeit infrequent firewood

# Apple

buds

bark-reddish-brown, with warty patches

½-inch diameter fruits

dark purple

# Black Cherry

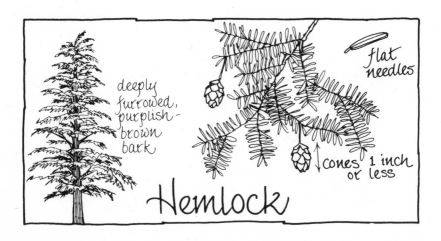

deeply furrowed, purplish-brown bark

flat needles

cones 1 inch or less

# Hemlock

Balsam Fir
*the fragrance of Christmas*

bark: reddish-brown scaly plates

upright cones

flat needles

White Pine

gray bark

needles: bundles of 5

white Spruce

brown bark

grows to 75 ft. tall

4 sided needles

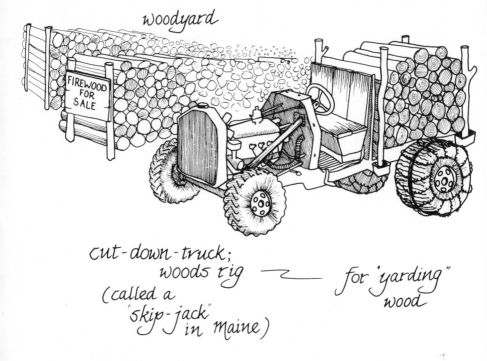

woodyard

FIREWOOD
FOR
SALE

cut-down-truck;
    woods rig — for "yarding"
(called a              wood
    "skip-jack"
        in Maine)

## Cutting Firewood

The traditional way of putting up cordwood is to cut a tree
down and into four-foot-long sections during the fall and winter,
leave the wood to age over the summer, then put it into the
woodshed the following fall. After trying to girdle trees and let
them age on the stump and also trying to cut trees during the
slack time of midsummer, I've returned pretty much to the old
way. The girdling (cutting a groove into the bark all around the
base of the trunk) doesn't always work and in any event the gir-
dled tree can take several years to die. Summer-cut trees are
still in leaf, making it hard to see limb location, so you can best
judge the direction of fall. Leaves are a nuisance when you are
trying to cut up limbs, too. They make stacking impossible, so
you have to come back after they've wilted and dried. So you
may as well do the whole thing after leaf fall.

### Equipment

The most essential piece of equipment in cutting wood is a
wood mover. If you are hauling sawlogs out of rough country,

the best thing is a team of draft horses. For cordwood, a farm tractor and wagon is good if you have one. If you don't, a good pickup truck will do for the wood and a lot of other hauling chores. Four-wheel drive is pretty essential, and heavy-duty suspension and cooling is good too.

A heavy-duty power winch attached to the frame, front or rear is good insurance if you get bogged in, but if you can afford it get a real (Warn or other manufacturer) winch, not one of those little half-ton capacity toys you see advertised in the discount stores and mail-order catalogs. For rough-country travel, see if you can't have a gas tank protection plate installed too. All power equipment that goes into the woods must have spark-arrest mufflers to prevent fires—by law in most places.

*Chain Saws*

Unless you have the ambition to fell with an ax and cut up your trees with a hand saw, you'll need a chain saw. Once again, avoid the heavily advertised lightweight saws, gas or electric. If there's one piece of advice I feel strongest about, it is the saw. Get the largest, most expensive model you can swing. Make it Homelite, Stihl, Poulans or any other brand, U.S. or imported, that makes saws for professional loggers. That little "Mighty-Midget" gas or electric you see for just under $100 in the discount store may be okay for pruning the lilac bushes, but it won't last through a cord of wood. Roller tips are nice; they let you keep a tighter chain but require added maintenance and a special grease gun, so the pros don't bother with them. Avoid saws with automatic chain sharpening gadgets; they don't work.

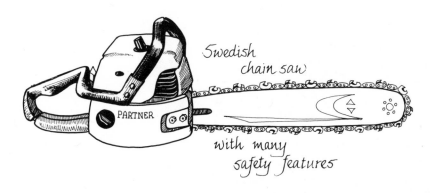

Swedish
chain saw

with many
safety features

*Maintenance*

When you buy the saw be sure to get three or four of the special round files for sharpening saw teeth and a small flat file for cutting down the steel nubbins—the spacers between each tooth—that kick sawdust out of the cut. The saw will come with a comprehensive instruction manual including routine safety tips. All I can add is, be sure to learn and follow the advice. A live chain saw is considered a deadly weapon in the logging country and just one slip can leave you short a leg. The main rule is to hold the grips tight and never push on the cutting bar. Let the saw do the work at its own pace. If it stops cutting, makes smoke or requires more pressure than perhaps a gentle rocking motion, your chain needs adjustment or sharpening. Any time you see burnt sap building on the top back side of the teeth, you should stop and go over the chain with the files. It is a good idea to give it a lick or two every 15 minutes anyway.

Saw noise is harsh, and prolonged exposure can damage your eardrums. Wherever you get your saw, you may be able to

*filing the teeth*

*Sharpening a chainsaw*

pick up some of the glass fiber stuffing professional loggers put in their ears; it only cuts out the harsh, painful and potentially dangerous wavelengths. Cotton isn't any good, but a set of ear-muffs will do double duty when it's cold. Look for Flents ear-plugs in the drugstore.

It isn't a bad idea to buy a small brush and use it to keep too much sawdust from clogging the cooling fins on the cylinder in the woods or the saw may overheat. Then when the cutting season is over, you'll want to clean the saw up completely and store it. Use high-pressure air or water to clean out the fins com-pletely. There will be a lot of oil and sawdust stuck on the engine and in various cracks and crannies, and you may have to spray on gasoline (outside, away from any fire, of course) before spraying or air-cleaning it. Then polish up the outside, empty the gas tank, start it and let all the gas run out of the carburetor. If it isn't running perfectly, tune it, by the way. Instructions are in the manual.

Then remove the spark plug, put a few drops of light engine oil in the cylinder, give the starting rope a few pulls to coat the inside of the motor with oil, and replace the plug. Put a light coating of oil all over the saw, the blade in particular. Finally, pump out a good supply of chain lubricating oil, and move the chain around by hand till the bar groove is filled with oil. Empty the chain oiler reservoir, pump out the little that remains in the pump, and put the saw away till next use.

## Felling

Although felling small cordwood trees isn't as dangerous as cutting down a yard-wide sawlog, any falling tree can kill you. You should know or decide where the tree is going to fall, and make sure it goes down there while you are a good tree-length away on the off side. Standing well away from the tree, hold a pencil at arms length, loosely so it dangles straight down. Com-pare the tree trunk with this and you'll see which way it leans. Put the pencil tip at the midpoint of the base and you'll be able to see which side is overheavy with limb growth. Walk around the tree, sighting past the pencil and compare lean with weight distribution. You'll probably make correct judgments your first few trees because you put the time into looking them over. It's later, when you've more experience and aren't bothering with

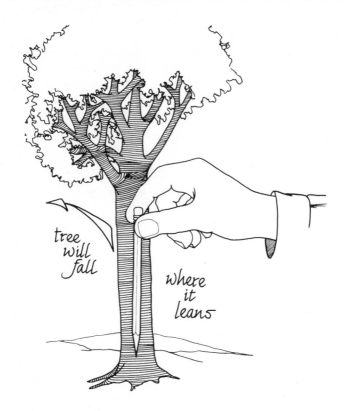

tree
will
fall

where
it
leans

the pencil anymore, that you'll make mistakes. Or that is how it has been with me.

You will surely want the tree to fall into a cleared area. If it goes down into more growth, it will likely hang up in the

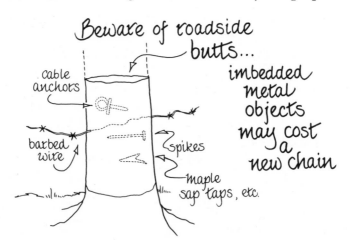

Beware of roadside
butts...

cable
anchors

barbed
wire

spikes

imbedded
metal
objects
may cost
a
new chain

maple
sap taps, etc.

branches of another tree, and a hung-up tree is a killer. The only thing worse is a widow-maker, a big—often decayed—limb that a falling tree knocks off itself or a nearby tree; the limb will fall straight down, never over with the tree, and more than one has lived up to its name. So first thing, plan your escape route. As soon as the tree begins to lean, get going. Always uphill or along

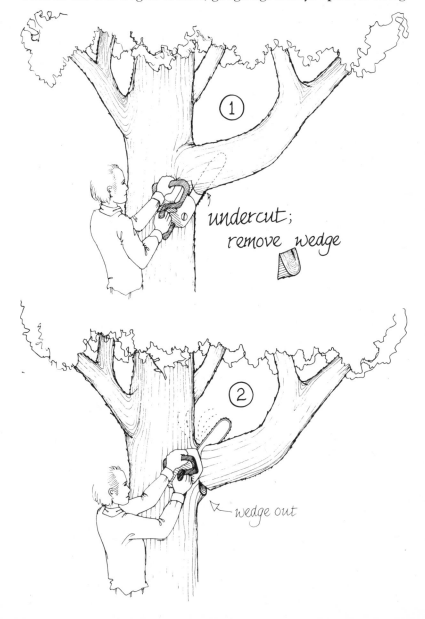

undercut;
remove wedge

wedge out

the slope if you are working on a hill. A tree can move downhill a lot faster than you can. I prefer to back into standing timber and have two or three trees between me and what's coming down.

If a limb or two appear to overweight the tree in the wrong direction, cut them off. Get up and secure in the tree, use the saw to make a cut about a quarter of the way through on the bottom of the limb, up close to the trunk. Then, lop her off by cutting down to the bottom cut from the top. Without the bottom

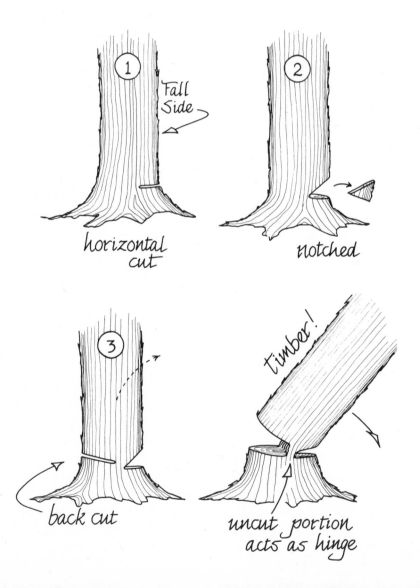

cut the limb may hang up on a hinge or take a great ragged strip of bark off with it as it goes.

## The Felling Cut

To fell, cut a notch in the fall side of the tree. Make it about a third of the way through. First, as close to the base as you can get, cut a horizontal lick. Be sure this cut is level and horizontal if you expect the tree to fall where you want it to. Then start an angling cut about a third as far above the horizontal cut as the tree is thick. Kick out the notch. Then to fell, go around the back and cut in horizontally a couple of inches above the horizontal portion of the notch. The tree should begin to fall—ever so slowly at first—when you are several inches from the inner face of the notch. The uncut portion will act as a hinge, guiding the tree down as you want it. If notch and cut were correctly done, though, the hinge should pop as the tree hits the ground.

When you sense that first perceptible movement, turn off the saw and hightail it away. The top can catch in another tree, a hidden irregularity or a hollow in the trunk can affect the fall and the trunk can swivel on the stump and the tree can do you in. I figure I have five seconds to get to safety. I don't know if they hand out awards for the longest distance run in a five-second dash, but if they did, a logger would hold the gold medal. And it's best to practice the dash every time you fell a tree, not on the very few occasions when you suddenly realize you must to stay alive.

## The Hung-Up Tree

If the tree settles down on the saw, binding it up, all you have to do is hammer in a wedge behind the bar. They sell soft metal and plastic ones, but we cut our own on the job when needed. Wedges can also be used to tip a tree slightly if you want to influence its direction of fall. Hammer a good wedge in behind the cutting bar just opposite the direction you want the tree to take or in the cut below a big limb that you suspect might pull the tree in the wrong direction.

Trees will hang up on you now and again, particularly in thick stands of small cordwood size. Resist the temptation to climb up, joggle it till it's free and then play Paul Bunyan and ride it down. If the butt is completely off the stump, you can use

## lopping off the limbs...

### ...close to trunk

**overhang:**
**cut from above**

**Section supported on either side:**

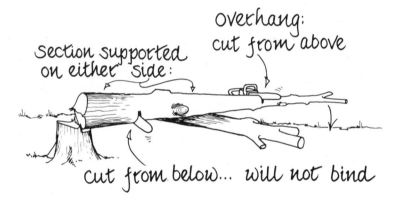

**cut from below... will not bind**

**limb directly under felled tree...**

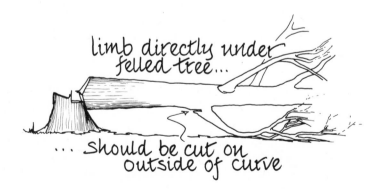

**... should be cut on outside of curve**

your peavey to try rolling the trunk around. If it is still on the stump, hinge unbroken or even just lodged up against the stump, there can be a tremendous amount of stress at that point. Pull it free and the butt can spring out like a battering ram, and pity the logger that happens to be in its path. It is dangerous to get in under a hung-up tree too. There's no telling when it will fall. I do one of two things. If possible—if there aren't stumps interfering—I loop a length of logging chain around the butt, attach it to the winch or rear hitch of the truck with stout ropes and pull it out. If that isn't possible, a fence stretcher will sometimes get it free. If that doesn't work, we go on to another stand and figure the winter winds and snow will do the job for us. Better, though, to plan each felling well and avoid the problem.

**Cutting Up the Felled Tree**

Once down, the tree gets another careful survey. The limbs it is resting on will be sprung, containing a great deal of compressed energy. Cut a limb under great pressure and it can break free and spear you. Or, if all the tree's weight is being held by a limb, and you cut it, the trunk can come at you quicker than you can blink. So be sure you know what is holding up what. And cut into the limb complex carefully. I like to commence at the top, the lightest wood, and trim away main trunk and limbs in short pieces—12 inches for kitchen range wood, 14 to 16 inches for the heating stoves. Trunks stay four-foot-long cord size over winter.

Cutting away a bit at a time, I section the tree as I go, working from the open end. By the time I get to limbs holding the heaviest trunk section, the great mass of the tree has been removed. If a limb does spring, the results won't amount to much because there isn't a great deal of tree attached anymore. If we have room to store them, small sticks are thrown into the truck as cut. Hauled back to the house, they get stacked for the quickest drying, under cover. If not, they just go into a loose pile on the ground. Air circulation won't be as thorough as if they were put in proper ricks, but the short length of each stick compensates. By next fall, all but the bottom sticks will be well dried. These will go on the bottom of the pile at the house, and by the time we get to them come spring, they'll be dried too.

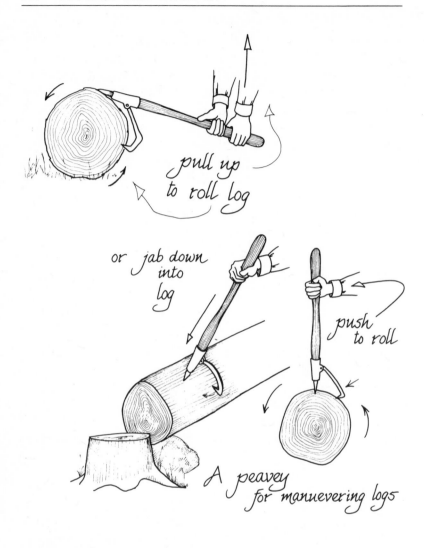

pull up
to roll log

or jab down
into
log

push
to roll

A peavey
for manuevering logs

The larger wood from trunks or the main limbs of the occasional large tree we take out are cut in four-foot lengths and stacked to dry over winter in the woods. Come the following fall, as we cut new wood for the next year, the year-old sticks are cut to length and split if much over six inches through. The easiest way to cut is to sink four or six posts into the ground in a rectangular crib three and a half feet long and as wide or deep as the cutting bar of your chain saw. (Six for cutting twelve-inch wood, four for two-foot lengths). Wood is stacked in the crib, and you just go down the stack, cutting several logs at a time.

log on Ground:
cut ⅔ through ... roll over to finish...

pulp hook

...or
prop up and saw through

in the winter woods,
brush away snow under cuts.

hidden rocks ruin chains

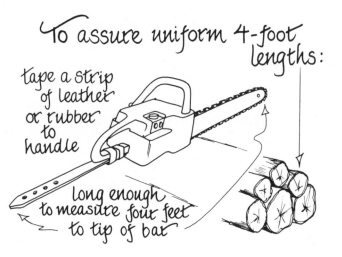

To assure uniform 4-foot lengths:

tape a strip of leather or rubber to handle

long enough to measure four feet to tip of bar

"Cutting Crib"

## Splitting

To split smallish logs, you can put one up on a stump, or hold it on the ground at an angle, resting against another log as in the photo, and clobber it with a splitting maul. (Wear stout shoes and safety goggles.) The maul is a variety of ax with a heavy, blunt and thick-edged head. Where a much thinner and lighter axhead would fail to split many sticks and lodge in the cut, the maul makes a wide, splitting cut.

For really large logs, you'll need splitting wedges too—like broad, thick axheads without handles. The first cut is made with a maul. You put a wedge or wedges in and hammer them home with a sledge. Generally, we split only very large logs in the woods, leaving most of it for winter exercise.

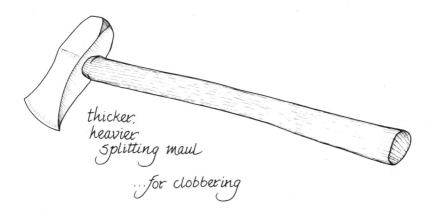

thicker,
heavier
splitting maul

...for clobbering

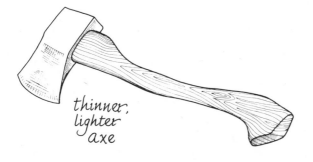

thinner,
lighter
axe

wide-splitting,
splitting maul

thin-cutting
axe

Splitting logs
for furnace

steel
wedge

Here's one way to split wood. Hold the stick on the ground with your best foot (the right, if you are right-handed). Have the splitting end—the upper third at least—resting on a half-split placed on the ground in front of your foot's position. Swing the maul so it is aimed to imbed in the ground log's forward curve—the leading edge—not in the top or back side. That way the blade can never get to your foot. If the blade slips out or twists off a thin side strip, nothing will get to you. Some folks don't agree with me on the safety of this technique. I prefer it to setting logs up on a stump, as illustrated, though. Splits don't fly the way I do it—an advantage if little kids are hanging around. But take care of your toes if you try it—wear stout boots.

At the end of a stroke, the maul should be in this position: well imbedded in the ground log, but far enough into the cutting log so that it pops in half. It takes some experience to know where to place the cutting log for best splitting. But be sure to move logs around rather than altering the stroke of the maul or ax. You don't want to lose a leg.

Here's how to split kindling. Use a good-sized hand ax. Hold a chunk of good-splitting wood, such as white birch, on a log and bring the ax down just hard enough to split out your sticks, pulling the holding hand away just before the ax hits. Too hard a whack and the stick may fly. Too soft and the log won't split and the ax will lodge in the wood. No need to elaborate on what might happen if you don't pull away the holding hand in time.

## Mechanical Woodbutchering Tools

While maul, sledge and wedges are the traditional splitting tools and pit or horizontal two-man crosscuts and bowsaws the log cutters, not all of us have the time and muscles of young Abe Lincoln (splitting off fence rails by hand, remember?). And many woodburners will have knotty chunks or such twisty-grained wood as elm or red oak rather than lumber-grade straight-grained trunk wood. For many people and much available firewood, some mechanical help in splitting is as welcome as the chain saw is in cutting.

The dandiest splitter I have ever seen is well over a hundred years old. It sits near an abandoned wood shed in the north country, right beside a yard-wide cordwood saw. Both were originally run off a steam engine; in later days off a giant spool fixed to the jacked-up rear axle of an ancient truck that is all rust but for a still-shiny enameled shield that reads Diamond Reo. With truck in high gear, the saw runs (ran that is) off a thick

leather belt actuated by an idler pulley. Four-foot cordwood went on the cradle to be sawn to length a truck load at a session.

Then the saw was stopped, belt removed, truck put into low and the splitter hooked up. It looks like a guillotine for giants— two ten-foot-high I beams a yard apart with a wedge on an H-shaped traveller that rode up and down in the greased channels of the inner edges of the I beams. They tell me that there used to be a variety of weights that could be hefted onto the wedge. As is, its quarter ton or so of weight should split granite. A flexible cable ran from an eye in the traveller through a pulley-block on the crossbar at the top of the vertical I beams and down to the revolving drum. The operator looped the cable around the drum several times, and much like a donkey winch on a ship, when he tightened the cable, it griped the drum, lifting the wedge. With finesse and experience, the wedge could be held more or less stationary at any height in the air—sort of half slipping. Still, it must have been a very trusting helper who rolled a log into position. The cable was then loosed, and gravity did the splitting.

The height the wedge was raised so it just split the log and didn't eat the old tire bumpers that caught the traveller and kept the wedge from splitting the upper peninsula of Michigan in half was a matter of judgment. Still, if done right, on an average log the cycling time it took for the wedge to go up and down was four, maybe five seconds. Primitive, some would call it, so dangerous to the uninitiated that Ralph Nader'd have it banned, but capable of splitting most anything that has a right going into a heating fire, and faster than anything on the market today. Also a work saver. Logs could be rolled onto the cradle from the lumber wagon, cut pieces rolled to the splitter bed, then split chunks would often as not fly out at right angles to the wedge and pile themselves at each side. Or so I'm told.

I don't mention this bit of ancient handmade ingenuity out of nostalgia. Like most nineteenth century industrial machinery, with exposed moving belts and such maimers as that wedge, it was a potentially lethal machine. But I do admire its simplicity, dual function, and efficiency and would like to have a miniature version. Here I'll come flat out and state that for our one-, sometimes two-family cordwood operation, I wouldn't purchase any log splitter or cordwood saw available today. (Purchase, that is. Renting is something else.) In my view there isn't a heavy duty woodbutcher available that's affordable for the family woodburner. None of us needs a commercial sawmill.

I've given thought to designing a small, compact and reasonably priced combination unit, and if it works out, I don't lose a hand in the process and Ralph Nader stays at home, I may try to sell you one some day. If you get it together first, let me know. (From time to time you do see ads for combination units, but all to date require a heavy farm tractor and use its three-point hitch, power takeoff and/or hydraulic system and cost as much as five years' supply of wood bought aged, cut and split and stacked in your woodshed.)

**Saws**

Now so far, we get by with a good chain saw and hand splitting tools. But, the two-cycle saw uses considerable gas and oil over a year's wood supply and its cut is twice that of a regular crosscut circular saw. In a lifetime of wood heating, that's a good lot of wood gone to sawdust, though whether it's better used in the stove or mulching the strawberries is a moot point. Still, a circular saw operating off a comparatively economical four-cycle

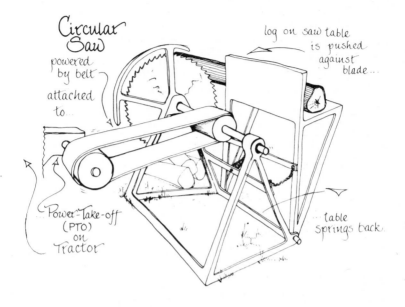

Circular Saw powered by belt attached to...

log on saw table is pushed against blade...

Power-Take-off (PTO) on Tractor

table springs back.

engine—particularly if designed to cut up smallish branches, maybe six inches through or less, which don't warrant the chain saw or are too whippy for the chain to cut easily—would be a big help in the woods and back home in the wood yard. Cutting them with a bucksaw takes more time and muscle than a lot of us care to expend.

The only small-scale circular saw you see very often is similar to the antique I found in the Michigan woods years ago, but smaller. The saw rests on a spindle which is supported on a metal sawhorse arrangement. The saw is powered by a belt attached to a farm tractor's PTO, and you can't find them much anywhere but on old farms. They are slow, dangerous and how many of us have a tractor to power one? (And if you do have a tractor, I can't help you find a new cordwood saw; the farm outlets I know that carried them a decade ago don't list them any more. Wait a bit; they'll be back.)

Homesteaders and large-scale gardeners are in somewhat better luck. Gravely, the firm that makes the justifiably well-known multiple-attachment walking and riding tractors, still sells a small log saw for its units. The blade isn't too big, but it will zip through branches faster than a bucksaw and will mow down small trees and undergrowth to boot. Except for Gravely, walking tractors were a thing of the past till the recent boom in gardening (brought about by many of the same factors that created the wood heat rebirth), and I note several multiple-use walk-behinds coming in via import, such as the fine quality BCS Mainline and several new domestic makes. All of these should be capable of powering a family-sized cordwood saw in addition to performing their gardening and other chores, and I'm keeping an eye out for the right one to come along.

## Splitters

The cheapest splitting device available is a miniature version of the guillotine arrangement described above. You put a log under the raised wedge and whack it with a sledge hammer till the log splits or the wedge becomes impossibly stuck in twist-grained wood. From what I've seen, any log this gadget can split easily will split with a single whack of a simple ten-pound maul. The maul will also do an easier job on twist-grained or knotty wood, as you can turn the blade in the cut, giving added split-

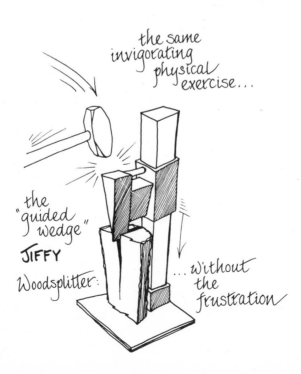

ting emphasis. Then when it sticks, a few jerks on the end of the handle will get it out for another crack.

Meanwhile, the mini-guillotine is imbedded in the wood and you have to try and dismantle the thing, knock around on the traveller to get the wedge out and finish up with the sledge or wedges anyway. These things cost the better half of a hundred dollar bill. I suspect that an enterprising timber country hardware dealer who offered a dollar trade-in for them on a good splitting maul would do a brisk business in scrap iron for a bit, plus he'd have the billet of wood each came stuck in to warm the store. Silliest thing I ever saw.

### Screw-Type Splitters

Not a bit silly, but very effective and scary as all get out are the threaded cone-shaped log splitters you've surely seen advertised as "the world's best log splitter" or some such similar thing. The principle is glorious in its simplicity. You simply jam a log at the sharp (and replaceable) end of the cone and the gadget literally threads its way into the wood. It is not fast—or isn't if operated at a safe rate, but will find a way to bust up the most knotty, gnarled and twisted chunk of wood you can aim at it.

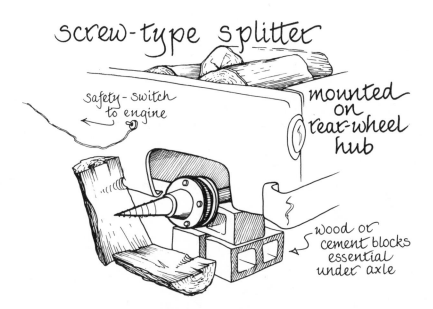

screw-type splitter

safety-switch to engine

mounted on rear-wheel hub

wood or cement blocks essential under axle

The problem is, the most heavily advertised version is the cone by itself; you are supposed to bolt it onto the jacked-up, left rear axle of your car or truck. The units come with an emergency cut-off switch for your auto engine (a bore to install, and most users won't bother). But suppose you thread on a stick, watch the splitting process begin—the splitting cone obtaining leverage the only way it can, unless you hold the stick and risk splinters flying through your skull when the stick pops, by pushing a log end against the ground. But you turn for another stick, the cone bites into an old maple sugaring spile or an especially tough knot, the log resists and the cone gives a great heave, pushing your car up and off the jack and. . . .

Well, it has happened, friends. And even if you do see it begin to happen, use the emergency switch, what do you do then with your car suspended on a half-split log, under bust-apart-any-second tension?

Good as the idea may be, I wouldn't put one on my truck if they gave it to me. If you have one and insist on using it, rest the axle on a real firm support such as a whole trunk load of cement blocks, carry a couple of heavy-duty screw jacks for good measure, stand clear, right by your emergency switch, when operating it. Then don't expect your auto dealer to stand behind any warranty or insurance agency to pay you for any damage done to your car. No vehicle was designed to be a log splitter and abuse such as this can produce damage ranging from bruised threads on the studs of your hub to a complete wipeout of drive train from axle to belly button.

The threaded cone costs something like a couple of hundred dollars at this writing, will likely drop in price as saner power sources are offered to make it go. A few firms are already coming out with complete units. I've seen ads for one powered by a four-cycle gas engine in the under-ten horsepower range and doubtless others are on the way. I've not seen any of these in operation so can't make an intelligent comment. However, a threaded device of any sort, a screw in fact, needs a lot of torque—a lot of lugging power—to work. Without getting technical, I'd be skeptical of the capacity of a purely mechanical transmission driven by a small garden-tractor-type 3500 rpm gas engine to put out that sort of oomph and hold together for long. Sorry to be so vague, but this equipment is just developing and like so much in wood heat has yet to stand the test of time.

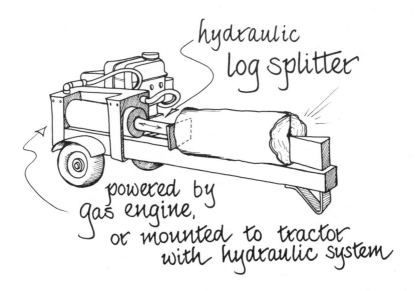

*hydraulic log splitter*

*powered by gas engine, or mounted to tractor with hydraulic system*

## Hydraulic Splitter

One proven mechanical log splitting device is the hydraulic logsplitter. The splitting maul and gravity splitters do their work by applying massed, moving energy—inertia—to the wood in one great whack that makes the opening entry and does the splitting in one operation. The cone-types enter by insinuation, then split by ever-widening the crack, backed up by a lot of torque. The hydraulic splitter uses simple brute strength, built up a dribble at a time in a ram.

As the illustration shows, the machine is a steel I beam with a wedge welded on one end and a hydraulic ram powered by a small gas engine at the other. You start the engine and put a log up against the splitting wedge. The engine runs a small pump. You open a valve that lets the pump move a light oil from a reservoir through super-strength tubing into the cylinder. Inside the cylinder is a piston, and as the fluid pressure builds up (fluid being incapable of compression, and the whole system built to hold thousands of pounds of pressure), the piston moves out, pushing the ram, which pushes the log. In time the log gives. With an adequately powered splitter, you need only pop good-splitting wood open, but with an especially stubborn hunk of elm, you may want to run the log all the way past the wedge. In either case, once the log is split, you reverse the valve handle and the pump sucks fluid out of the cylinder, pulling back piston and ram. Cycle time on a good splitter with a two-way cylinder, operated with skill, is five to ten seconds per log. You can split a lot of wood in a day's work with it.

A good splitter comes on wheels with a trailer hitch, road license and lights, has an industrial-rated gas engine and a hydraulic unit that puts out 15,000 pounds or more of pressure per square inch and cycles fast. It also costs well over a thousand dollars. Fine for a commercial woodlot operator, but on the rare occasions I need one, we rent it for a half day.

Hydraulic systems are found on every car, tractor and airplane made, and on much industrial gear, so components are easy to come by new or used. And of late we are seeing any number of small shops offering complete splitters and mail-order outfits selling building plans plus major parts. A number use little gas engines and relatively small hydraulic units, say a three-and-a-half-horsepower engine and a cylinder with a stroke of eight inches or so. This is fine if well-made. A small unit can build up as much pressure as a big one, only it takes longer. The problem is, with the small splitters I've seen, you take more time to get the job done than with a maul. You have to lift the log onto the splitter, actuate the pump, wait a half a minute or so for the pressure to build up, remove the log and finish the partial split by hand. All the while, you could be splitting a half dozen logs with your maul. I'm afraid that I have to put the little splitters in the expensive toy category; you will have at least four hundred dollars tied up in the smallest unit. If you've got the money and maybe a bad back, fine. I don't have either.

You may be tempted to pick up a set of plans to build your own, have some fun and save some money. It isn't all that complicated. You can have the welding done at a job shop, scrounge wheels and axle from an old trailer and put together a hydraulic unit from parts picked up at junkyards, farm equipment outlets or wherever. However, you have to know what you are doing. Hydraulics is a technical field all its own. The pressures needed to split wood are tremendous and you can't slap together just any combination of pump, cylinder and hoses and have it work. The inlet and outlet couplings where pipes or hoses run between each segment of the system have to match—and you'll find hydraulic connectors come in all kinds of different fitting, thread and material variations. It isn't a job for the average handyman.

# Cleanup

If you've ever been in the logging country of Maine or the Northwest you've seen the mess a conventional logging operation makes of the woods—tractor ruts everywhere and slash (all but the main log of each tree), lying wherever it fell. I like to leave our woods as orderly as they were before we went logging. We cut up and use all branches that are much over an inch thick—they make great kindling or biscuit wood for the range. But there is still a lot of little stuff remaining. In cuttings located close to the road, I haul our shredder/grinder out from the truck and shred the slash, then haul it home for garden mulch. Far back in the woods, we pile it more or less where the crown of each tree falls as we cut through the wood. There's no particular reason for doing it, I guess, except out of respect for the woods and the wildlife that will find winter shelter in the piles. A series of little brush piles looks tidier than a general mess. If the trees and rabbits don't care, we do.

You may not agree that a tree deserves respect. But the Indians thought so. They populated every one with a spirit and apologized for cutting it down. Louise and I don't go quite that far, but we can't put out of mind the fact that every tree is the product of so much of nature's time and energy. Especially the real old-timers. Take that big rogue white pine I mentioned earlier. Why, it must be 250 years old. The trunk is a good five feet through at the base and then some. If trees could think, it certainly would remember the Nipmunk tribe and their campground across the east branch of the Swift (you can see the location from the tree's upper branches), or the yahoos that set out from Boston in the early 1700s to massacre those gentle people and establish Volunteersville. Or Daniel Shays and his farmer's rebellion that ended in a massacre of another sort when the army regulars hunted them down in the woods just a few miles away. Or the peaceful years when its trunk was surrounded by "mowin' fields" back when Nichewaug was prosperous and John Carter's shutter mill buzzed away just downstream. The old tree has outlasted at least two barbed wire fences that we know of. The rusted ends of the wires twist out two and three inches deep into its bark, one behind the other, at the same level on the downhill side. It has Nipmunk arrowheads buried deep in its heartwood too; I'm as sure of that as I can be of anything. And since it's the right distance from the road (in sight but out of hearing) it likely hides a few overgrown heart-and-arrow-and-initial carvings from turn-of-the-century wanderers away from the hayride. It weathered the hurricane of 1938 without a scar. That was the storm that levelled half the county and split one of our maples so it's shaped like a giant slingshot. The tree even survived Clarence Hunt. Its big south-growing limb is where the deer hung that he sold (quarters, halves, whole with hide on or off) before they put in a hunting season and bag limits, and for some time afterwards. And then in his later years, that's where he'd nail up the stuffed head of a 12-point buck on opening day of deer week, then sit in the shade, rocking and waiting, to whoop uproariously when a city dude would saunter by on the road, steadily peering into the woods, then do a double take, fumble at his safety and let go with both barrels into the tree trunk. Hell, it even survived me back when we farmed not far distant. I learned the hard way then that you can't tie up a newly electric fence-wise hog by its neck, even for a few hours. It was that tree I tied her to, and it was one big hunk of that tree's root I had to hack out to make burying space for three hundred pounds of self-strangled, unbled and thus inedible, pork.

I don't advocate any romantic "poems are made by fools like me, but only God can make a tree" nonsense. But, doggone it, that tree has been a living presence through most every day of recent local history. It's survived it all and doubtless will outlast me. The tree has seniority, and we nod formally when we pass on the cart road.

## CHAPTER SEVEN
# Rettin' Up Come Spring

Once the weather has warmed enough in Spring that the big stove is let go cold but for an occasional open fire to take the chill off a rainy evening, it's time to begin the spring cleaning inside and out. Rettin' up, my grandparents called it. And until you've lived in a wood-heated environment you can't know what an old-time spring cleaning was all about. In squeaky-clean modern homes the automatic baseboard heat and self-cleaning ovens deserve little more than an annual dusting. But with wood heat you have a house that needs an application of strong lye soap and a lot of elbow grease.

You see, each time you open a fire door or lift a stove lid, a bit of smoke and a gentle rain of fly ash is let loose. You'll get more smoke lighting up under a cold flue, puffs of smoke and sprays of ash during exceptionally windy weather, when someone opens the front door suddenly, and when adding wood or stirring up the fire. Of course, you keep ahead of the black cobwebs that seem to grow every night, and you dust the wood or painted flat surfaces. But at the end of a heating season the windows will be covered with a bluish film, and there likely will be a nice round smoke stain on the ceiling over each fireplace and stove and the same on the wall behind. Slipcovers (or the throws Louise keeps over winter on the upholstered but non-slipcovered furniture) need a good pounding on the rocks; so do the rugs, drapes and curtains.

We've found that the main offender is the light ash that seems to fall continually on everything. It can make a proper mess if you go at it with soap and water, but on nonfabric surfaces, it comes off completely with a light dusting or vacuuming—that is *light;* it will get ground on if you push too hard. A little vacuum cleaner with a soft brush on the hose seems to

easy-
wipe
surfaces...

save
hours
of
fly-ash
cleaning

work best. Windows throughout the house and light-colored woodwork in rooms near the stove will need a good cleaning with soap or detergent, though.

## Decorating a Wood-Heated Home

In a wood-heated home you quickly come to recognize the advantages of a lot of great-grandmother's decorating ideas, long since branded as old-fashioned. Throw rugs can be hauled outdoors and beaten. They waxed all their floors in the old days too—easier to clean. Modern wall-to-wall carpets are a disaster with wood heat. Modern kitchens have spice racks, dish shelves and all open to the air, which grandmother insisted should as much as possible go into the closed pantry or be hidden behind tight, slickly painted doors. It's a lot easier to wipe off a painted cabinet door than to have to wash every dish before each meal. They tell me that old-time waxy shelf paper has just about disappeared from the supermarket shelves along with the thumb tacks to hold it in place. Well, we've rustled up a good supply, and I'll bet you will too. There just isn't any way you can keep that layer of ash dust from settling on every flat surface, and it is convenient to be able to remove the old shelf paper (and the layer of tinfoil kept on top of the range, et cetera) and replace it at the end of each heating season.

One of grandma's conventions that we have abandoned, though, is white lace curtains. Our house came equipped with a set of turn-of-the-century curtains, complete with a battery of wood and brass-brad curtain stretchers. After the first week or so of woodburning, we came to suspect the reason for those curtains; they are probably the most efficient dust and smoke filters known. Inside of a month the white curtains throughout the house were filthy, while walls and furniture seemed fine. It must be that warm air circulating through the filmy lace carries the dust to be trapped, thus preventing it from landing anywhere else. Cleanliness of the white curtains was surely a status symbol in the old days, and if granny hauled the curtains out to the washtub every week or so, I can see why she welcomed the oil stoves. Down our curtains came and out they went. We find that print curtains and dark or mottled burlap or other heavy drapes do a better job of insulating on cold days, and if they collect dust it isn't noticeable till after the whole lot of them has returned from the cleaners in May.

**Cleaning Tips**

Now, there's no secret to washing drapes or floors; that's done in even the niftiest of modern no-work homes. But there are a few cleaning and keeping-clean tricks we've learned by trial and error. First, as intimated above, you should keep as much of your house as possible slick and hard on the surface. Then the ash just wipes off. Anything that isn't so surfaced should be hidden behind closed doors or be easily removable. Remember grandma's stair runners? They were strips of dark, hard-finish carpet kept in place with a few carpet tacks—easily removed for a good scrubbing outdoors. Today's light-colored shag stair ruggery stapled down for good would turn your stomach after a year of woodburning.

Home-furnishing designers, relieved of the need to produce readily cleaned goods, are turning to unfinished wood, as in butcher-block tables, and all sorts of matte-finished items, particularly in plastic, where removing the normal glossy finish somehow connotes quality. Well, any grainy, porous material just begs to be clogged up with fly ash. We had a lovely little rough-wood table that slowly turned black over the winter despite all we did. Washing in bleach only raised the grain and produced splinters, which only gathered more dust. An application of polyurethane varnish helped somewhat, but we finally gave up and replaced it with a smooth, well-polished piece.

Ash collectors they didn't have to bother with in the original wood-heated days were lamp shades. Chimneys on oil or kerosene lamps had to be cleaned of lampblack every few days anyway, and what shades they had were easily cleaned brass or tinware. The white wire-frame and nubbly fabric shades we have today are the next-best things to dust mops, but we've never heard of a method of cleaning them. The foundation of many shades is paper of one sort or another; water and soap would rust the wiring, damage the paper and shrink the fabric. We asked a drycleaner if he could clean lamp shades and he just blinked. So we learned to do it ourselves. You need a good-sized washtub and enough nonflammable cleaning fluid to fill it to a depth of a couple of inches. Then the shade is swished back and forth rapidly, revolved slowly till clean, then shaken and let dry. Ultimately, we plan to shift to easily cleanable shades of stained glass, perforated sheet metal, perhaps oilskin. In time we'd like to have a house that you can wipe clean with a dust cloth. Likely never will get there.

## Spring Cleaning and Maintenance

House cleaning really doesn't take all that much time, but it is hard work; rather than force it, we wait for one of those lovely early spring days when you can open up all the doors and windows and go fishing or plant the spring garden. But don't you give in yet—we try not to. If you let the spring chores go by the boards, you are likely to forget them altogether during the sun months. Washing drapes is depressing in October when you are firing up the stoves that will get the drapes sooty all over again. Worse, if you delay the other aspects of spring cleanup, you may fire up a creosote-clogged stovepipe or flue and burn your house down. So we tackle the chores as a family and normally get them done over a two-day period.

### Flue Cleaning

Usually, Louise and the kids start upstairs. There are winter clothes and toys to sort out, bedding to shuffle and the scrubbing to do. I'm relegated to the roof and lower stories to clean out stoves and smoke pipes. First it's out with the flue-cleaning gear. Heavily used woodburning flues simply must be scoured at least once a year. Tire chains, as mentioned earlier when we were reconditioning an old flue, will work. There are several patented flue scrapers advertised that don't look like much to me. Fortunately a few brush makers are making flue brushes generally available—try a hardware or other store that sells a good assortment of wood stoves. Our own flue brush is an oldie, meant originally for scrubbing out coal furnaces. It's a foot-plus long, with stiff wire bristles bound up in twisted cable, and it works. There's a loop at the tip of the brush; to this I fasten a weight—a large coffee can filled with rocks, then a watery cement slurry and an eyebolt running all the way through. A length of strong synthetic rope is tied to the eye, run through the loop in the brush, then half-hitched up the center wire spine to the end of the brush handle. With newspapers and masking tape, I seal up every place the flue opens into the house, lest soot blow out over everything. Then it's up to the roof, where I run the big old brush up and down till the beam of the big lantern shows all loose creosote has fallen down. Then I shovel out the cleanouts, remove seals and forget the flues for a while.

making a weight with eyebolt.

cement binder optional

juice can.

and rocks

*Brushing the Flue*

This big old brush is about worn out and its type isn't made anymore to my knowledge. In the near future we'll have to make do with standard flue brushes, as will you, most likely. You need one to fit the precise size and shape of your flue; they come round or square, in the size of any standard flue tile. The wire bristles should be a bit bigger around than the flue so they will press tight against it, but not so big that the brush can get stuck. In any event, you will find the brush has (should have) loops at each end of the central spine. You'll need two people and two lengths of rope as long as the flue is high.

Fit snap-rings or some sort of clip to the rope ends to hook onto the brush. Then one person goes up on the roof and lowers a rope down the flue. The other grabs the rope through the

nylon rope
half-hitched
along
flue-brush

and
tied
to
weight

cleanout, hooks the top of the brush to the rope, hooks his rope to the bottom and inserts the brush into the flue. The roof top worker hauls the brush up the flue, detaches it and tosses it back down for another go around—as many as needed to clean the flue. The helper's rope is insurance; you can usually pull a jammed brush back down, though the arm work through the confines of a flue clean-out is brutal. The moral is not to get too big a modern flue brush. Once started up, the bristles will all be aiming down and retrieving them from the bottom is something like trying to pull an open umbrella down the flue by the handle; you have to reverse the attitude of the thing and that isn't easy.

With insulated-metal flues be sure to have a right-sized brush. I wouldn't even use a standard flue brush, but would shop around for one with soft bristles—well soft by flue standards—of cane or rough fiber. This is particularly true if you

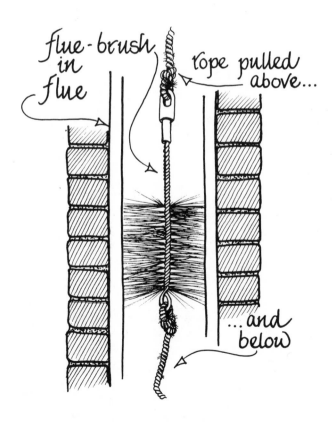

flue-brush in flue

rope pulled above...

...and below

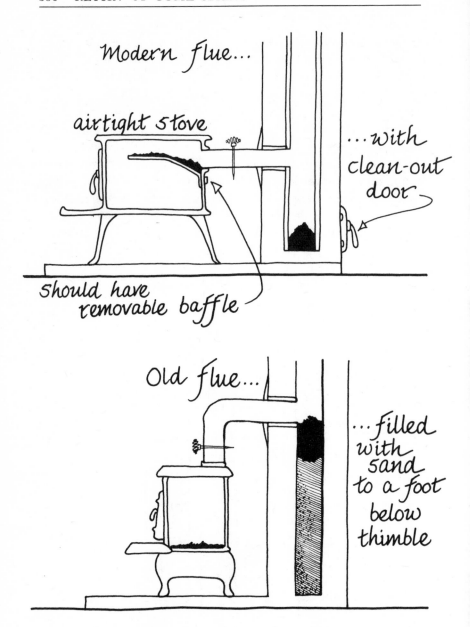

Modern flue...

airtight stove

...with clean-out door

should have removable baffle

Old flue...

...filled with sand to a foot below thimble

have bends in the flue. You don't want to have to dismantle the whole stack to retrieve a stuck flue brush.

Some older flues will be odd-shaped. Old-time chimney sweeps used brushes on long poles like giant toothbrushes to scour all sides of the chimney. (Or they lowered little kids into the flue to do the work.) You won't want to use either of those methods. Tire chains are what I've used. Some people make a special flue cleaner of scrub brushes mounted on a wooden frame which should work. And finally, for problem flues or roofs that are higher than your ridge-running experience considers safe, you may be able to find a chimney sweep.

### Chimney Sweeps

For years we country woodburners were our own flue cleaners, else the flues didn't get cleaned. But with the rebirth of general interest in wood heat has come the renaissance of an ancient trade—the chimney sweep. It's a good thing too, as wood stoves are turning up in homes of a lot of people who lack time, equipment or desire to go clambering around on roof tops,

a chimney sweep

to say nothing of the knowledge of flue systems needed to do a good job. Problem is, neither do a goodly number of people who call themselves sweeps.

Just as anyone who paints WOOD STOVES on a storefront sign can pass himself off a wood heat expert to the unwitting, anyone with an old top hat and tails, a stove brush and a length of rope can claim to be a chimney sweep. Who's to know the difference? Well, you, for one. First, there are a number of chimney sweep schools popping up. They are passing around sweep skills that have been kept alive in Europe. A sweep with a certificate from a school at least should know the top of a flue from its bottom. Don't let the top hat and tails schtick sway you, though. It is the traditional costume of European sweeps—people who in the old days were considered the absolute pits in the social order and begged the cast-off clothing of morticians, who weren't held in much esteem either.

Today, of course, the costume is just that and kind of fun. Still, it's no guarantee of a competent sweep. Telephone your

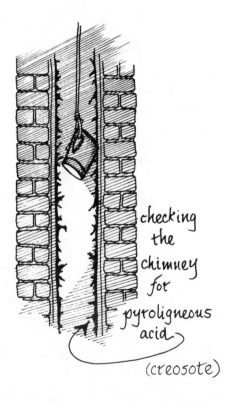

checking the chimney for pyroligneous acid (creosote)

potential sweep and ask what his or her (sure, there are women doing the work) routine is. First should come a thorough inspection of the exposed part of chimney, wherever it is exposed, outside the house or inside (inside being most important). Then all fireplaces and other openings should be sealed up tight. Proper cleaning gear should be determined by the in-house inspection, but once on the roof with it, the roof-top member of the team should lower his cleaning rope with a brilliant lantern attached to check the inside of the flue all the way down (the first cleaning anyway). One otherwise competent crew skipped this step on one new flue I know of and the brush got lodged where the flue connector jutted out a bit into the liner. They also forgot to hook on the helper's rope, and the brush is still there, awaiting spring and the arrival of the mason who built the obstruction in the first place. (And I oughtta know; it's our own freestanding flue. I'm fine on a safe roof, but don't like dangling out in space three stories up, so called in some sweeps. Made the mistake of assuming they knew what they were doing—I made the same mistake about the mason too. Well, I hope you benefit by our misfortune. No, I'm not kidding, but then I never did claim to know everything about wood heat and don't you trust anyone who does.)

Anyway . . . once inspected and sealed, the flue should be brushed by at least a two-person crew, the soot removed, fireplace, if any, cleaned out with a big industrial vacuum, and you should get any report of flue damage found. Don't expect the sweeps to clean out your ashes or your stovepipes unless you pay extra. Around here the fee is about $50 for the first flue, somewhat less for additional flues on the same roof. That is a high price to pay for what is really semi-skilled labor performed by part-timers (I don't know of any full-time sweep), but competition is already bringing the price down. And the number of flues going up is increasing the available work daily. In time North America will once again have a corps of experienced wood heat people who will offer a full line of services, including flue cleaning. They will be full time, have insurance against stuck brushes and such mishaps and will qualify as genuine experts. For now, in chimney sweeping as in most other aspects of wood heat, you are well advised to become your own expert, even if you have someone else do the actual work.

## Summerizing the Stoves

Stoves need some attention now. For a good cleaning, all removable parts such as baffles or inner liners should be removed, ash vacuumed out of crannies and the whole thing scraped, brushed and vacuumed again. Check all bolts, tightening or replacing any that are loose. Small cracks between poorly mated or warped castings can be patched with stove cement. Any major cracks in the castings or steel plates are a sign of major trouble. You can try to have steel cracks brazed shut, though no patches will hold up for long. A new plate can be cut and welded in or you may decide to change stove brands. Warped or cracked plates in cast-iron stoves can sometimes be replaced. You will have to break the stove down, removing all bolts and cleaning out inter-joint stove cement. If the stove was well made in the first place and has not been overfired or used long enough to warp the plates, a replacement may fit. You'll probably have to use a good bit of cement, and be sure to watch the stove carefully during the beginning of the next heating season. If the stove opens up, replace it.

Before retiring a stove for the warm months, polish the exterior well. Enameled stoves need only a good wipedown. Plain black iron or steel needs a good coat of stove polish to protect against rust. Paste silicone stove black is considered better than the liquid version; real stove freaks rub it in thoroughly with their hands, then polish with a soft cloth. It washes off your hands with a little paint thinner.

And, finally, if you have an airtight stove with asbestos gaskets in the door or around cooking lids, plan to replace them, every year if you use the stove for serious heat. We find that the grinding action of ash continually getting into the gasket plus plain wear and tear simply uses up a gasket in about a heating season and a half. So a fresh gasket every spring is a good way to assure that your stove will be a real airtight come the next fall. If the stove shop who sold you the unit doesn't stock gasket material, raise cain. Matter of fact, be sure they have the stuff before you buy the stove.

You'll find that the gasket material is sort of greasy, very flexible stuff. You can chisel out the old material and just push the new into the channel. Close the door tight overnight to set the gasket. Unless you want a real chore next year, don't cement the gasket in—no need. In laying it in, pack it close, don't stretch it out. It will dry out and shrink somewhat in use, and you want plenty of gasket in the door to effect the seal.

Any door latches or other moving parts should be lubricated. Don't use machine oil. It will burn and actually clog things up. A silicone grease has a mineral base and won't burn on you.

## Cleaning the Stovepipes

Once flues and heaters are clean, I dismantle both stovepipes. Pipe goes outdoors to be scoured clean, but before cleaning, each joint is inspected for airtightness. Inspect each joint carefully; if you see little tongues of soot extending out on the male—the crimped—end of the joint, you've got an air leak letting smoke out into the room. If, after dismantling the joint, you find clean, unsooted-up tongues on the female—noncrimped outer section of the joint—the leak is letting room air into the pipe. The former leak can be dangerous; the latter can affect draft. You'll find leaks most commonly around the seam in one or the other pipe. Seldom can you cure them by putting in sheet metal screws. Better is to smear stove cement on the crimped end at each joint and hammer the pipe together well. The cement will break for the next cleaning, but you may have to knock the joint around a bit to break the seal.

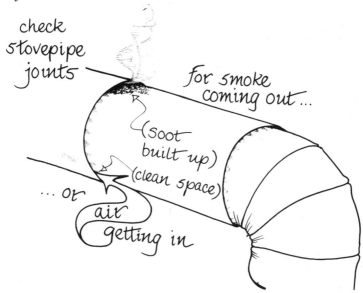

check stovepipe joints — for smoke coming out...

(soot built up)

(clean space)

... or air getting in

*Reading What the Creosote Says*

Before you take a brush to your stovepipe or the flue connector, take a good look at what's inside. You can expect the pipe to be coated with a thin layer of soot all around, probably with a windrow of fallen soot and some ash in a pile at the bottom. If the accumulated creosote at the pipe bottom is excessive—more than a quarter inch deep, say—you may have been burning green wood, been operating an airtight in closed-down mode too much, or you just may have a combination of stove and flue that generates a lot of crud. Resolve to clean out the pipe and check the flue for creosote buildup several times during the year.

Now if the inside of your stovepipe looks like a miniature version of Mammoth Cave, but all in black, with great sheets and curtains of creosote buildup dangling from the pipe, you almost certainly have a poor joint between stove and flue. Particularly if you are burning an airtight stove, you *must* assure a perfectly airtight seal between the air intake in your stove and the top of the flue. The most common leak occurs where the pipe fastens to the stove itself. This is especially true with imported stoves; their smoke holes are sized in metric measurement, while commonly available North American pipe comes in inches. A lot of people just tack a too-loose inch-sized pipe on their Lange or Morso or other imported stove and hope for the best. Well, with a loose joint, the flue draft pulls in cool room air just where the smoke is exiting the stove. The constant chill hits the smoke, condensing the volatiles, which rapidly accumulate on the pipe walls in swirling, eddylike sheets—the sheets actually showing the pattern of smoke movement as interrupted by the infiltration of cool air. You simply must seal up the joint. Best is to get proper sized pipe and hammer it on for a good friction fit. Alternatively, you can coat the stove outlet with furnace cement and push the pipe on. Or, you can put on a seal of fireclay, push the pipe on and hope the clay, which will dry out but not fire rockhard, will make a good seal.

If your pipe is normal, but the unusual creosote buildup occurs in the flue connector or in the lower part of the flue itself, your leak exists where the pipe enters the flue system. We have the potential for such a leak with our living room heater. I just stuff fiberglass in around the pipe where it enters the wall at the thimble. This must be done each time the pipe is cleaned, but it is essential.

**Cleaning the Range**

Earlier I recounted the fun of cleaning the inside of a wood range. Well, the outside is a joy in its own right. Unless you have one of the cubelike modern designs, you'll be faced with a whole network of iron or porcelain filigree, bumps and grooves, all of which were made to catch and hold grease spatters, which in turn are among the best ash traps there are. So once a year, the entire behemoth gets a scouring with hot water, detergent, steel wool and a strong alkali oven cleaner for the sticking places. Fortunately, the one part of a good wood range that does not need oven cleaner is the oven. Modern gas or electric ovens are heated by burners or elements at top or bottom, and any splashes stick and harden on the walls. The wood oven is warmed by heat radiating from all sides. The steel or iron liner gets hotter than the oven interior, and it is heated whenever the stove is used as a space heater. Any spatters turn to char and just slough off in time. I guess a wood range has the *original* self-cleaning oven.

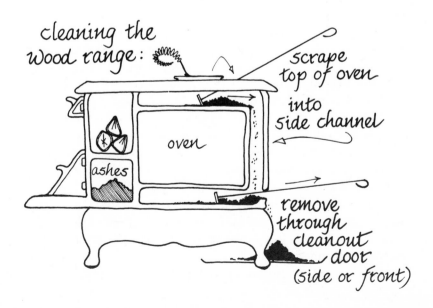

cleaning the wood range:

scrape top of oven

into side channel

oven

ashes

remove through cleanout door (side or front)

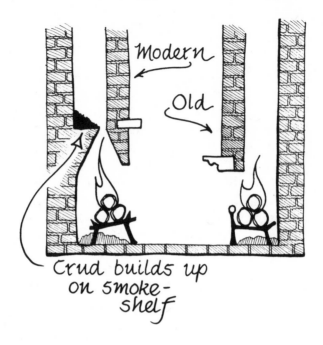

Modern

Old

Crud builds up
on smoke-
shelf

### . . . And the Fireplace

If you're using a fireplace for serious wood heat, you have a couple of special cleaning problems (problems, I fear, that have not been considered by the makers of several new gadgets that supposedly "turn your fireplace into a stove"). Particularly if you vent an efficient airtight stove into a fireplace flue, you'll have a significant creosote buildup. It simply must be removed, *and* when you scour it out, the majority is going to fall down and accumulate on your smoke shelf. Now, how to get to it?

Here lies one of the beauties of David Howard's angle-iron damper, illustrated in the fireplace chapter. You can pull it right out and get your hand or a brush (perhaps an old dish or toilet brush bent to a 45 or 90 degree angle) through the throat to scrape out the creosote.

But what if you have a hinged damper? First, be sure the damper is fully open before you start cleaning, particularly if it is the sort that hinges up and back to rest on the floor of the smoke shelf. Close it while flue-scraping and the buildup of soot on the damper plate and the shelf behind may actually prevent you from opening it, to say nothing of getting the creosote out. I'll

cleaning creosote

out of smoke-shelf

with bent-handled brush

say it again, modern fireplaces and fireplace flues weren't made for serious heating, and cleaning them hasn't entered into the thinking of the housing industry for a hundred years and more. So, in converting a modern, decorative fireplace into an effective heater, keep the cleanup in mind. I'd say, go through the mental exercise of the annual or more frequent cleaning before you so much as order a stove to fit a fireplace flue—and certainly before you select a damper when having a fireplace built. I've seen some inventions that slip into a fireplace and attach to a central hot air or hot water heating system. They have pumps or blowers and need all sorts of electrical wiring. But some of them are designed to utterly block access to the flue. To clean up, you'd have to dismantle the entire assembly, drain pipes, break solder joints and unwire everything. It would take an electrician, a plumbing contractor and the town building inspector just to get you access to your creosote buildup. Pretty silly, I'd say. So, again use your common sense in selecting your wood heat gear. And don't forget the cleaning job that comes up each spring.

## Woodchopper's Cleanup

Once the house is shined up well enough to please most any mother-in-law, I turn to the equipment and wood storage areas. First the edges of splitting wedges, maul, ax and hatchet are filed to a proper sharpness. Ax and hatchet should taper to a thin point, and the sharper the edge the easier the job. I use a set of hand files; a medium file to remove significant amounts of metal, a fine one to hone the edge. The old-time loggers who supposedly shaved with their axes would hone them with an oilstone till they were sharp as knives. That's going a bit far I'd say, but a sharp edge does make for better cutting. If you have an old pedal-operated oil- or waterstone, it will do a lovely job on a cutting edge. But never use a high-speed stone. It will heat up your edge and remove the temper.

You want relatively blunt edges on your maul and wedges—the better to pop a stick apart. But I like to file the cutting edge as sharp as I can to give the steel a better entry. A wedge in particular needs a good biting edge; with a blunt edge it's hard to get a wedge started in a log. Be sure to grind off any "mushroom cap" burrs on the blunt end of your wedges. These can break off and take out a good piece of hide as they fly.

Once all edges are honed properly, the metal is scoured with steel wool to remove any rust or chipping paint, then all blades receive a light coating of machine oil. Next come the handles. Each is checked for tightness, and I make sure the wedge that holds the end of the handle in the head has not begun to work out. If it shows signs of protruding, I pound it back in and let the head and handle sit in a bucket of water overnight.

All wood is checked for splinters. Any place where grain has raised is pared with a sharp knife and sanded lightly; then the wood receives three applications of raw linseed oil, applied several days apart and rubbed in well—rubbed till the wood is good and hot. Then the tools go down into the cellar where they will remain cool and relatively moist through the summer. If they were just tossed in the rear of the woodshed, the summer heat would dry the wood out to the point that the handles might be ruined.

## The Woodshed

Then there's the woodshed itself. By spring the wood supply is reduced to a pile of oddsized sticks. They are hauled off and stacked loosely under the porch. There they will dry out over the summer and provide good kindling come fall. What remains are piles of sawdust, dried out bark, mouse and bumblebee nests and assorted twigs and chips. If left all summer, this can dry down to the point that it is positively dangerous. "Tinder dry" is more than an expression; more than one woodcutter has gone out to his woodshed, stirred up a cloud of wood dust, lit his pipe and gotten blown sky-high in a genuine explosion.

More common is the danger of using a littered woodshed as a work shed over summer. Most homeowners have powered lawn or garden tools, which leak oil or gasoline, paint cans, brushes soaking in solvent, and all sorts of flammables that if stored in a woodshed can soak into the wood debris without being noticed. Spontaneous combustion is a danger, and so are sparks from any number of sources. So all the junk is shoveled up; it makes a great mulch for fruit trees, berries or flower beds. Then the shed is hosed down completely and it's ready for summer duty.

# CHAPTER EIGHT

# The Fringe Benefits of Wood Heat

Most folks with the good sense to heat with wood will also have a big garden and may even make their own soap. We do both, and the "waste" from the fires, wood ash, is a big help in each chore.

## Wood Ash

You'll recall that the ash contains all the minerals that went into the wood, the water and carbohydrates having gone up the flues in the process of combustion. These minerals were brought up from the subsoil by the tree, whose root structure is almost a mirror image of the top growth, going down many feet. The nu-

Spreading ashes on the winter garden

wind

trients the roots bring up are the tree's gift to the topsoil and to you and me, who live on the produce grown there. To rebury a tree's final gift to us in some dump is a waste, almost an insult to the plant that has warmed us and cooked dinner. Ashes should be used.

## For the Garden

Pure wood ash contains three garden helpers, in varying amounts depending on the wood. But a good hardwood ash will contain something under five percent of the plant food potassium in the form of potassium carbonate. There is also about two percent of the plant food phosphorus in the form of phosphoric acid. The potassium is what makes strong stems; the phosphorus makes for better roots. These two elements together with the other mineral salts in the ash are alkaline in nature and like lime, they will tend to sweeten your soil. Most garden soils are more acid than most food and ornamental plants prefer, so the addition of wood ash improves the soil's pH—the scale used to measure acidity/alkalinity of soils. Most garden plants prefer a pH of 6.0 to 7.0, which is nearly neutral. You'd be best advised to make a soil test to see where your land fits on the pH scale. It is possible to oversweeten the garden.

However, you'll get 50 to 60 pounds of ash from a cord of wood, and most gardens will benefit by having that amount applied each year to each five hundred square feet, a plot a bit more than 20 feet on a side. You'll find that hot ashes sprinkled on the garden in the cool early spring will help keep down the weeds at the time they are most vigorous. If you've problems with slugs or snails, sprinkle a generous layer of ash at the base of affected plants. The slimy little critters don't like to crawl over the dry ashes. Ashes sprinkled liberally over the crowns of root crops—onions, carrots, beets—will prevent egg-laying by the females of many species of "root maggot" fly. The maggots are larvae of bugs that gnaw around in the vegetables. The damage they do is often minor in itself, but their tunnels admit disease that can wipe out an entire crop. We never worry about early spring hatches of maggots; if anything they help thin the carrot rows. Besides, ash must be dry and unleached to be effective, and the early spring rains make the ash treatment impractical. But once things dry up in early July, ash is kept in the root crop

rows. Plants are mature enough that the mild alkalinity can't harm them and the bugs have to go elsewhere to plant their broods.

We've found that ashes are an effective barrier to all sorts of insect pests. A ditch around a garden section, some four to six inches deep and filled with ashes, keeps cutworms out in the sod where they can't get to the young tomato seedlings. To date

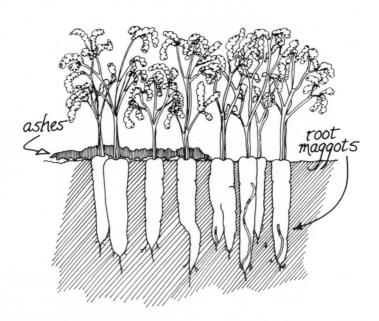

ash-filled
ditch dug
around
tomatoes...

...to keep out
cutworms

we've had no problems with bark borers in the young pit fruit trees—peach, plum and apricot. These are larvae of insects that lay eggs at the base of the tree. After hatching the little worm crawls up the trunk and burrows into a tender branch. A circle of ash is maintained around each tree; not only does it seem to deter the borers, it keeps sod from growing up to the tree's trunk and competing for water and plant food. Dry ash is powdery, slightly astringent and has an unfamiliar smell. I'm convinced that it is an effective deterrent to small, low-to-the-

circle of ash
around
fruit trees...

... to deter bark-borers

ground garden marauders like rabbits, though I'd not trust it to protect the cabbages against groundhogs or the young corn from raccoons.

**For Making Soap**

The combination of potassium and the other caustic salts in wood ash is most commonly called potash. "Pot ash" is wood lye, which when mixed with animal fats, becomes soap. Grandma's lye soap. The early woodcutters turned much of our eastern forests into soapmaking potash for sale to Europe, a criminal waste by today's standards. But we can do the same with ashes left from our own heating fires.

*saponification*

*Making Lye*

First, you want to be sure to use only pure wood ash. Ashes from newspapers, magazines or other paper waste often contain clay and other additives that you may not want in the wash. You'll want to sift out any unburned chunks, though there should be precious few if you've kept a well-made heating fire. You can make a simple sifter with a square frame covered on one side with hardware cloth and a broomstick for a handle. Just sift out each collection of ashes, and pack the fine ash away till you've a rain barrel full. A big bakery-sized covered flour sifter would work for ashes too, if you have one.

While rummaging around in the old brick ash pit in our cellar, we found a Triumph Ash Sifter. Made around the turn of the century by the Success Manufacturing Company of Gloucester, Massachusetts, it contains a revolving screen and does a dandy job of sifting ashes, keeping most of it contained so you don't get ash dust all over, as you do with a hand sifter. I doubt you'll find one for sale anywhere, though it would be easy to make one like it if you've got a knack with tinsnips.

We just set the sifter over a plastic trash barrel with a lot of small holes punched in the bottom and sift until it is full. Then the barrel is set over a washtub and soft water is poured through it—rainwater being the easiest to get. We keep adding sifted ash as it compacts, pour more water on every now and again, then let the leachings evaporate naturally. What's left is potash (lye), and you'll get up to five pounds per barrel. The caked residue in the barrel still has some garden value, though most of the potassium is leached out. So what's left goes on the land. We do make it a point to burn only apple wood for a few weeks now

and then. After a particularly careful sifting, leaching and re-washing, the apple wood ash is given a second sifting through a fine screen. Louise mixes the ash in water and uses it as a glaze for her stoneware pottery.

## Then Soap

When the potash liquid is pretty well evaporated, we find friends for the kids to spend the day with and make soap. The lye is highly caustic, and children shouldn't be permitted to get anywhere near it. In the morning we fire up the outside fire-place and get a strong cooking fire established. One six-quart pot is filled with potash. It is boiled slowly till all water is gone. We then dry bake it till all the carbon has burned off and the potash has toasted to a light gray powder. It should be stirred during the smoky, final dry-baking stage. Next we weigh out one-pound portions for each batch of soap.

At the back of the cookstove Louise keeps a one-pound cof-fee can with an old-fashioned fine-mesh tea strainer over it. Into the can goes all the cooking fat from sausage and bacon, roasts, chicken; any animal fat or mixture of fats will make good soap. Each can, not quite filled to the top, will hold two pounds of fat, and we seem to accumulate over four cans each year. More can be produced by frying up beef tallow. You can buy unsalted lard—expensive, but highly refined—which makes lovely soap. On the farm we always saved fat from slaughter of all ani-mals, freezing it till we had enough to justify starting up a wood fire outside under the big old, round-bottomed, cast-iron rendering pot we found in the barn.

In any event, waste fat must be cleaned; for each batch of soap we put two cans of fat into a pot half-full of water, stir and boil for a half hour. The pot is let cool a bit. Then we strain the warm fat carefully through cheesecloth to remove any foreign particles.

Now, all we need is a third iron pot big enough to hold fat, water and lye without boiling over, two ladles and a long-han-dled spoon to stir with. Cast-iron pots are best, though porce-lain-lined ware is alright. Never use aluminum pots to make soap. The 12-quart stainless steel canner is of sufficient size to hold our recipe of four pounds of fat plus a pound of potash dissolved in three gallons of water.

Saltless
Soft
Soap

## Saponification

The soapmaking process is called saponification. I vaguely recall studying it in high school chemistry. How it works I can't recall, but if you combine water and fat or oil with any one of several caustic compounds, it all combines to form soap. Just dump them together in a barrel and you get soft yellow soap that will eat holes in the wood in time and will do likewise to your clothes or the family dog that gets washed in it. But cook the mixture in saltwater and you'll get a nice, white hard soap that is mild enough to use for bathing if it is cured for six months and you've a tough hide. All green soap is caustic at the start and the longer it ages the milder it becomes.

There are several good books out on soapmaking now, and in them you'll find a lot of recipes and good directions. But here is the old-fashioned three-pot boiled soap we learned to make back before the books were available. First, a pound of potash is stirred slowly into a gallon of cold water in one Dutch oven; this will heat up and can burn, so the emphasis is on slow stirring.

Lye water
and
hot fat...
...dribbled
into
glop
pot

The lye water is kept hot while a gallon of water is brought to a boil in the canner and four pounds of fat and a cup of water is heated to the point where the water is bubbling up through the fat. Then as Louise stirs, I dribble in ladles full of lye water and fat till all is mixed.

The glop is let simmer for the afternoon, four or five hours. It should be stirred constantly, but seldom is. Then another gallon of water is brought to a boil, a cup of salt is added and this is slowly added to the mix. All is stirred constantly and cooked slowly till the soap forms on the surface. If you've ever let a bar of soap sit overnight in a soap dish full of water, you recall the ropy, slimy stuff that resulted. That's what new soap looks like. Scoop some out and if the fat and water don't separate, and the mass holds together, you have soap. It is still caustic, and will remain so for several weeks, so it should be handled with rubber gloves. We line a pair of two-foot-square wooden trays that are about four inches deep with cheesecloth. Then the soap is ladled in and covered to cool slowly. The leftover saltwater is poured into a hole in the ground and buried. Soap will harden slowly from the bottom up. Usually there is a thin layer on top that won't harden completely; it is scraped off and discarded. In

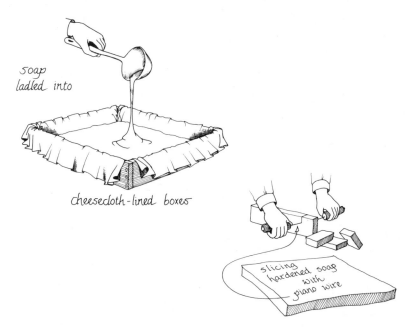

soap
ladled into

cheesecloth-lined boxes

slicing
hardened soap
with
piano wire

about a week the soap will be hard enough to be cut into bars; it
is worked out of the trays with the cheesecloth and sliced with a
length of fine piano wire with wooden handles at each end.
From the four pounds of fat, one pound of potash and some 20
pints of water we get about 50 standard "personal-sized" bars.
They are stacked on their ends back in the trays so air will flow
around them freely and they will cure completely, as mentioned
earlier, for as much as six months.

Once cured the soap can be used for about any home cleaning use. Chipping it up into soap flakes on a cabbage grater is a
lot of work. But for washing dishes we cut bars to fit an old-
fashioned soap-saver, a little screen cage on a handle that
Grandma used to hold scraps of soap. Swished through hot
dishwater, it works just fine. For dishes or laundry, you can
make a semi-liquid soap. Just chip a bar or two into several
chunks and drop into a jar full of hot water. In a few days they
will look much as they did when first scooped off the boiling pot.
The liquid soap does well with hand washing or the old-style
drum and wringer washers. But if you've an automatic washer,
don't try homemade soap as modern cleaning technology can't
handle the old-fashioned suds from soap you've made for free
out of the "wastes" of your wood heater.

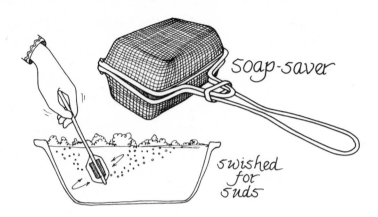

## Soot

As someone has probably said, if you like the ashes, you'll love the soot (from cleaning the stovepipes, flues and chimney cleanouts). It contains many of the minerals found in wood ash plus about two percent nitrogen. Florists use it in potting soil and so should you. We save up soot and add it to the mixture of garden loam, peat moss, sand and lime that we use in the spring to start the tomatoes and other long-season garden plants. Mixed with milled sphagnum moss and varying amounts of loam and lime, depending on the plant's requirements, it is also a good addition to greenhouse or houseplant growing medium. It makes any soil a deep, rich black color. It also smells to high heavens, so let it age so the tars can dry out before using it inside the house.

Soot also goes by the names of lampblack and carbon black and is good for any application where a dark coloring is wanted. Masons use it to darken up mortar. It's a good pigment for paint and varnish. Make your own ink by boiling the soot up in a little soapy water. It writes best with a quill pen. For one of those you need the biggest bird you can find; take out a wing or tail feather, whittle the tip off at a slant and split the tip. It takes a lot of dripping and shaking but is a good way to write letters by the light of your wood fire on a winter's evening.

## An Old-Fashioned Wood-Heated Wash Day

To get the most washing power out of your soap you'll want lots of piping-hot water. And all the old-fashioned soaps we've

*Washday team*

made somehow seem to suds up best in an old-time wash tub. If your cookstove has a water front, you are in the scrubbing business. All you need besides soap, your brush, a wash board and lots of elbow grease are a couple of buckets of water, a good fire in the stove, an hour's time and a big dipper. A kettle of water will heat up on the stove top in about the same time.

Our great-grandmothers had a little laundry stove with a small firebox and a wide top. On it went the oval tin "washer" with wood handles at the end in which she heated the wash water. Often lye soap was chipped in and the wash was literally boiled clean. Many old laundry stoves have special holders around the sides for flatirons—different sizes for different chores. You can heat them up on the cook top too. Grandmother's starched ruffles, lace curtains and other items needing ironing have pretty much gone by the board in today's drip-dry society, but in ironing out your sewing and such, an old flatiron beats any modern, electric, lightweight steam iron. They are heavy, and once hoisted up on the ironing board, do the work for you as long as you keep the bottom waxed (with wax paper). Just keep one or two heating while another is at work. You'll

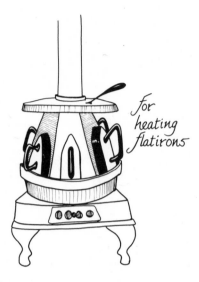

for
heating
flatirons

scorch more than one piece of cloth in learning how hot to heat them.

And when it's bath time for the children, out comes the wash tub, and one big kettle of water per bath is heated up. The kitchen becomes the bathroom in winter, of course. The big

clip-on
wooden handle
always
stays cool

a "rotating"
irons

no
cord
to snag!

kitchen range is well fired and with the oven door open, it's the warmest place in the house. And after each small ear has been scrubbed pink, the kids are dried with towels that were warmed up on the stove's racks—and the towels go right back to dry overnight.

## Psychological Value of Wood Heat

Finally, let's not overlook the intangible benefits of wood heat. Thousands of people have been forced into makeshift shelters during the power outages that are becoming common during our increasingly unpredictable and brutal winters. With a wood stove or two, they needn't have suffered so much. I mentioned a roughly similar situation that faced my own family shortly after we moved into our present home. Just the big kitchen range was sufficient to keep the house liveable, pipes unfrozen, and the family well fed till the emergency was over. You'd be surprised at the feeling of security and independence wood heat can give you. Blackouts, brownouts, fossil fuel crises, OPEC and Exxon, government energy rationers, blunderers and inepts can go hang for all a wood burner cares.

Think how utterly dependent a conventional North American home is on just one utility-provided power supply: electricity. Most conventional home-heating plants, a large proportion of kitchen appliances, your light, home entertainment appliances, most hot water, the water pump if you live in the country, and on and on are totally dependent on those electrical wires and the increasingly expensive juice they bring into your home. Comes a failure and the woodburner is just barely fazed, while his neighbor may be in real trouble. A good wood stove or two can keep you warm and fed, heat your dish and bath water and melt snow if the pump is off. About all it can't do is bring in a TV program, and that's just as well. Because if you do have to rely on this pretechnological energy supply for any length of time, you'll find that the elbow grease involved in life before electric appliances provides about all the entertainment you need.

I'm sure that in Granddad's boyhood he saw little benefit to the work of hauling in the wood, while his father must have disliked the splitting, and his mother the cleaning of ashes off every flat surface each week. They had no choice, and we can't begrudge them all the sigh of relief that accompanied the kerosene stove into the kitchen. But we and our kids (you and yours too, I hope) are returning to the hard work of wood heat willingly, even enthusiastically. I guess we welcome the environmental benefits and freedom from Exxon and its sisters that our willing wood labor brings as much as an earlier generation welcomed the relief from forced labor brought by the kerosene range.

And the exercise you get in splitting and hauling the fuel supply is made somehow more enjoyable by the close contact it gives with the wood. Wood is a bit of nature, a reminder that this log once was a tree in a forest, home to squirrels, birds, perhaps a raccoon family. The feel of wood is good. The roughness of the bark, the occasional splinter you have to dig out of a thumb. I don't even grumble (much) when cleaning up the bits of bark leaves and footprints that always leave a trail after the small helpers and I have brought in the day's wood supply.

Perhaps the prime attraction of the fireplace or stove is the fact of having a central "hot spot" in your home. The romantic symbolism of the fire in the hearth, with its connections to Mom and home cookin', is obvious and best left to more poetic souls

baby sleeps where it's warmest

register

than I. But life in a wood-heated home has more variety, more texture and novelty than living in one of your modern split-levels with electric, gas or oil heat. They are probably more uniformly comfortable, but typify to me at least the bland uniformity that seems to be overpowering life these days. With a wood fire, you can go out to the woodshed and split wood in below zero weather, then come back in and toast your snowy mittens to steaming right in front of the blaze. The fact that you can makes the chopping and the cold not only bearable, but downright enjoyable. With a centralized heat source, your whole house will exhibit a variety of climates for a variety of activities and people. Near the fire is hot and dry and just right for Grandmother's rocker. The warming shelf above the wood range is grand for rising bread or sleeping cats and the house dog sleeps warm underneath the oven. In the sitting room, facing the fire but over next to the window is where my reading chair goes; I need the slight chill to keep from dozing and can always pull closer to the heat if need be.

Except for the baby, whose crib goes over the register in the floor of the bedroom just over the kitchen range, people sleep best in cold temperatures. With plenty of blankets, of course. But the chill air coming in from outside carries moisture—good to keep your head and chest from drying out and leading to colds and flu as in many conventionally heated houses. If the cold is too much, pile on more blankets or open the door more or cut another register in the floor.

And, speaking of baby—once he or she puts on a few more years—what's more down a child's line of work than helping bring in the wood? Not that our preschoolers are assigned chores just yet. But a little guy enjoys being able to pitch in with the real work around the place when the mood strikes, and bringing in the logs (he's gotta try the biggest) now and again, gathering up sticks in the woods, and even driving Daddy half nuts by insisting on hanging around to fetch flying sticks when logs are being split, is a real contribution. In a wood-heated house a child—even a very small one—can participate in the real, meaningful task of keeping the whole family warm and well fed. The wood he hauls in is a genuine help and he knows it. It is not make-work or pure play and it's more than several cuts above sitting slack-jawed in front of the TV all afternoon. Which says something good about this (and maybe other) aspects of life at a pretechnology level. Everyone can help, really help, even the smallest members of the family.

And I could go on with wood heat's extra benefits. The cellar is unheated and chilly all year long—fine as is for me if I'm planing down boards for a new table, but when Louise comes down to make a few pots, and the children move into the playroom converted from the old coal bin, it's good to fire up the instant-heating sheet metal stove and warm things as long as needed. Same with the stoves in my little den in the attic, the workshop in the barn and so on. With wood heat you can be just as warm as you want, when you want, where you want. And you find that when you have this flexibility, this variety, you take advantage of it, adjusting heat to activity. The constant change does you good, keeps your head clear and your eyes wide open.

So we can even add a bit of psychological advantage to the physical benefits the exercise of chopping and splitting gives you, the money you can save, the valuable fossil resources you avoid using and the general environmental benefit of burning a clean and self-renewing fuel. Anyone who hasn't changed to wood heat and can, should. We hope you do if you haven't already and that this book helps a little.

Stay warm.

# APPENDIX A
# Wood Heat's Future

I'm amused to read the original text of this appendix, written just a few years ago. Just about all we predicted—such as domestic manufacture of Scandinavian-type stoves, increased demand and supply along with lower prices for combination fuel furnaces—has happened. None of us had any idea how fast wood heat would boom.

I admit to a certain fad influence in wood stove popularity, but if there was ever any doubt that wood heat has a solid, promising long term future, recent events have dispelled it for good. Already a good 40 percent of the families in Vermont have permanently retrofitted their heating systems to use wood as the main or supplementary fuel. A check of the small town newspapers in our part of central Massachusetts shows that the vast majority of building permits issued are for installation of wood stoves, each a major investment, in heat—not for fun. And it's the same all across the continent; anywhere the weather gets cool and there are trees, people are heating seriously with wood.

Domestic heat is my prime interest, and yours too most likely. But industry is beginning to look to the woods for more than wood pulp and lumber. One of Vermont's big electric power generators is using wood on a part-time basis and that state and others are planning all-wood burning power plants. The buzz word in industrial use is "wood chips." I've seen the machines they use to gather wood on the scale needed by a big plant. Huge grapplers go in, rip out whole trees, then jam them into a giant shredder-grinder that spews the chips into big trucks. The chips can be compressed for greater efficiency in hauling. The chips are fed to a modified coal furnace where controlled airflow and powered smoke recycling gear can assure near complete combustion. The federal energy people and a lot of industry big shots are looking seriously at wood as fuel, and I wouldn't be surprised if the minicar you buy a decade from now will be made from machines powered by wood.

There's also a good chance that it will contain wood-derived plastic parts and run on methanol—alcohol brewed from wood. After all, coal and petroleum are nothing but ancient trees that have been fiddled with a bit by nature and time. With appropriate technology and investment, the U.S. and Canadian economies could literally run on wood—indeed I'm sure will be doing so in part long before the petroleum runs out in a generation or so.

The danger, of course, as when industry gets its greedy hams on any resource, is that it will be plundered for a quick profit. If past history holds any lessons for us, it is fairly certain that much of our forestland will be misused. In the view of many, much is already being handled improperly by clear-cutting paper companies. However, even clear-cut land can be replanted. And that's the most promising thing that wood heat holds for our future. Trees are self-renewing—are the only burnable fuel source that is. And as long as you and I and other wood-conscious folks keep the politicians from letting the big energy companies turn the national forests to a desert, our grandchildren will stay warm and have jobs and live well. Not to say that a wood-powered economy can produce as opulent a lifestyle as we've had during our brief energy binge with petroleum, but we North Americans can continue to praise our luck at hatching out on a continent so rich in resources.

The total wood/solar life is a generation and more away, so what can we woodburners expect in the reasonable future? Regulation for one thing. We've mentioned that several states are beginning to mumble about defining just what constitutes an honest cord of wood, that building codes are beginning to move in on wood heat installations. That's just the beginning. Look for the beady eyes of government regulators everywhere. Any sector of economic activity that grows as significant as wood heat is going to attract some bureaucrat looking for a new field to call his own. Already in our locale there's a lively competition for authority in wood stove design and installation approval springing up among the Aggy Extension Service people who think they inherited the turf because wood comes from trees, which are plants, and they do forest as well as farm products, and the building code people who feel that stoves are theirs because they go in buildings, and the heating equipment regulators who have an engineer who needs work and thinks he can tell good stoves from bad. Then there are the insurance companies and their allies who think stoves should all be approved by Underwriter's Labs.

All of these influences, of course, tend to concentrate power into the hands of the big stove manufacturers or sellers with the political influence to "advise" regulators just as lobbyists do in all levels of government. These are also the folks with the money to pay for whatever laboratory tests the regulators decide on. I'm told that cost of tests in one lab runs over $50,000, not the sort of thing to attract small companies. And from what I've seen, the testing labs don't know anything about wood stoves despite all the engineers on their staffs. One tester turned down a fine line of Scandinavian stoves, reportedly because they felt the baffle systems cooled the smoke and the gaskets in the doors could wear out. Well, cold smoke and worn gaskets can cause problems for sure, but (probably unbeknownst to the testing people) they are problems that can be avoided by a knowledgeable operator. I mean, you can burn your house down if you leave a hot UL-approved electric iron sitting on the ironing board, can't you?

Now, I don't advocate unsafe design, installation and operation of any woodburning device. And frankly, given the great numbers of newcomers to wood heat who are running stoves these days, I'd like to see a tough set of code regs, adopted and enforced in every community. But this whole business of testing and approving—even the bickering over who is going to test and approve what—smacks of a political catfight, the kind that produces the typical governmental mishmash of overlapping, conflicting and generally incomprehensible rules. If my guess is right, the rule makers are just as ignorant of the realities of wood heat as I was when I nearly burned our farm down, and some of the rules are going to provide some amusing reading.

The trend in product regulation these days seems to be make everything idiotproof. Much of the regulation, of course, is a great benefit to the consumer. But if they decide to idiotproof wood heating appliances, they'll turn them into automatic devices not much better than an oil-fired central furnace. People rebelled against those nasty seatbelt buzzers that they stuck on cars not long ago. And if the regulators automate the charm, self-reliance and pretechnological simplicity out of wood heat, there's going to be an uproar. From this quarter, at least.

Actually, I'm sure that whatever rules are finally implemented will make sense in the long run. But I look forward to a shakedown period among regulators and regulations, just as there is bound to be among the growing numbers of stove and other wood heat implement designers and manufacturers.

As for heating machinery itself, governments permitting, we are bound to see continued innovation in domestic design and more and more good imported ideas coming in. The Scandinavian dominance among quality iron stoves is being solidly challenged by both domestic manufacture and by imports from other areas of Europe. Several French and Belgian stoves are new to our market, quite sophisticated in concept and time-proven. I'm sure we'll see continued inputs from the European stovemen; after all, they kept the wood fires burning while we gave in to oil, and they won't hesitate to apply their know-how to our comparatively huge market. I look for a new wave of imported stoves to challenge the original Scandinavian designs.

Another import of sorts that is making a minor stir in wood-

### An old Russian Stove
(one of a variety of flue arrangements)

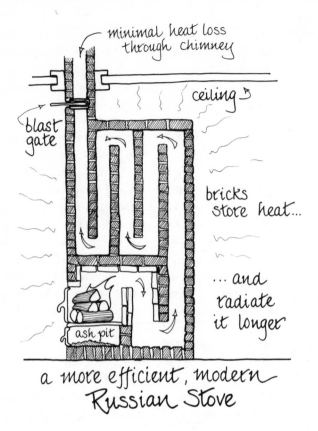

minimal heat loss
through chimney

ceiling ⤴

blast
gate

bricks
store heat...

... and
radiate
it longer

ash pit

a more efficient, modern
Russian Stove

burning circles is the Russian stove. This device, as Liz's illustration shows, is a huge masonry device that you would literally build your house around. The secret is that great heat sink; you build a fire—a great blazer—once every day or two, and the masonry holds the heat, radiating it slowly, in between fires. The "blaster" located up near ceiling level provides oxygen for secondary combustion in the smoke, and some users claim that their stack gases are nonexistent. Hmmm . . . Before building one, I'd be darned sure I could get in and clean out the smoke channels. But here is another good old idea that a lot of people will be looking into.

Here are a few other ideas—novelties now but perhaps a feature of every modern home in another generation.

Some years ago, while traveling in the Orient at Uncle Sam's request, I ran across a few old but good wood heating and cooking ideas that folks might be able to use back here in the petro-

leum-short future. The Koreans, for example, dig a firepit or
build a stove at one end of a house and exhaust a large cookstove
through a series of tile pipes that run under the floor and connect with a flue at the far side. An underground version of the
extended stovepipe. I understand that the ancient Romans did

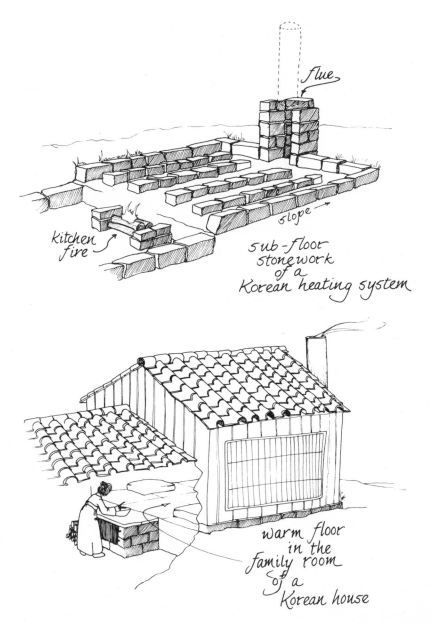

flue

kitchen
fire

slope →

sub-floor
stonework
of a
Korean heating system

warm floor
in the
family room
of a
Korean house

much the same thing to heat up their communal baths. The whole system would have to be fully airtight to keep carbon dioxide out of the house, and it would take a pretty tall flue to draw heat and smoke under a large house. You'd also want a fireproof floor; the Roman baths were solid ceramic. I would imagine you'd have to do a good bit of experimenting to arrange a series of dampers or other means of adjusting air flow through the system. And how the subfloor tubes are cleaned, I don't know. Brushes on really long handles, I'd guess.

Before their flower culture became an Americanized Chevy culture (or Datsun culture) the Japanese cooked and heated with unvented braziers, shallow pots containing glowing charcoal. The kitchen fire was kept in a series of firepots that resembled built-in hibachis. Rooms were heated with movable firepots, and the most ingenious application I've seen is the pit table. There is a round hole in the floor just knee-to-heel deep. Under the table in the center is the charcoal brazier. Diners sit on the floor, feet in the pit with a large tablecloth over the table and their laps. The fire warms feet and legs and a warm robe plus hot tea or saki warms the rest.

One warning, though. The houses where this was done were anything but airtight; the carbon monoxide that charcoal produces in great quantity just rose and drifted harmlessly out under the eaves. More than one family in a modern weather-

ventilated food warmer

Japanese pit table

firebox with draft holes

proof home has suffocated as they slept because the fire in the charcoal barbecue that cooked dinner was left to burn itself out inside. Charcoal uses a suprising amount of oxygen, even if you don't see much flame. So don't you dare use it in the Japanese fashion unless your home is thoroughly ventilated. This goes for any other kind of fire too. I recall one of the back-to-the-land publications publishing plans for a sawdust stove that reportedly didn't have to be vented since the sawdust fire can be made almost smokeless. Now this is fine out in the open, and I've seen such stoves keeping things more or less comfortable for winter work in the mostly open-walled cutting shed of a sawmill. But inside a closed space it would be a killer. Indeed, the more smokeless and cleaner-burning the fuel, the more dangerous the fire because you won't have the smell of smoke to warn you of danger. Be on the safe side and vent any wood fire. A couple of windows open part way at the top will do for cooking indoors on a brazier (which is what any outdoor charcoal cooker is). I'd never let an unvented fire burn overnight, no matter how many windows are left open.

One of many developments from Maine is being tested by the state's Audubon Society. The specially designed superinsulated furnace is a true "downdrafter;" the fire practically burns

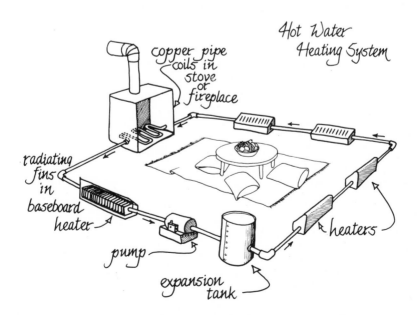

Hot Water Heating System

copper pipe coils in stove or fireplace

radiating fins in baseboard heater

pump

expansion tank

heaters

upside down as a baffle system directs smoke back up through the coals. Then the stack gases are passed through a gravel bed. Most of the heat is absorbed, stored in the gravel to be drawn off by the building's central heat unit (air or water pumped through the warm gravel just as it is run through a furnace heat exchanger in a conventional central system). This gravel bed also stores heat collected by a battery of solar panels. Quite a system, and if you'd like to keep track of it and other developments from a state that is one of the last bastions of wood heat in an economy that's pretty well given in to the oil companies, join the Maine Audubon Society, Portland, ME 04101. The membership fee is nominal and their publications well worth the small cost.

The combination of wood and solar heat is just beginning to be explored. No solar unit I've seen is really effective after two or three days of heavy cloud. Well, comes the overcast, you fire up the big Russian stove or light up a space heater or two.

And by the time you read this, there will have been further developments. The Maine Audubon system may not be feasible for most people's homes, but it's just a matter of time till someone comes out with a unit combining the best of the old and new ideas that will be priced like any other home heater. And the home inventors haven't abandoned this technology that we all can understand either. Our wood heat scrapbook of magazine and newspaper clippings on innovative ideas keeps getting thicker.

As I see it, innovation will increasingly stress fuel efficiency. We will begin seeing reproductions of the fine old stoves of the 1890s that turned all the wood into heat, not creosote or cheery flame. Why? Growing expertise and acknowledgement of the sophistication that can be built into a woodburner for one thing. But cost of fuel will enter in too. Already cordwood has doubled in price since we first took up serious wood heat. Some of this increase is traceable to inflation and higher fuel prices—for the logging machinery. But most of it is due to simple demand. More people with more fires means more demand for wood, and if the woodcutters can get a better price for their goods, they will take it.

How high wood will go in relation to other energy sources no one can say. The higher it goes in sheer dollar terms, though, the more people will be attracted to the trade. I suspect we'll see a cyclical movement in price. No one knows just now how many cords are stashed out in the woods curing right now, to come on the market next fall. I'd bet that high prices the fall of one year will get the woodsmen out into the timber to cut more than the market needs the coming year, so we'll see prices fall. But those low prices will reduce supply, which will increase prices again. If you've the space, you might stockpile a couple year's wood supply in the cheap years. Get cured wood for the coming heating season, cheaper green wood for next year. It will be ready, but not rotten and gone punky, when you need to use it.

As prices rise, more and more woodburners will be taking to the woods themselves. Wood heat types are doers by and large, and comes the time a cord of purchased wood nears the price of 150 or 200 gallons of heating oil, a lot of folks will invest in that chain saw and go lumberjacking. Indeed, weekend woodcutting is already becoming something of an acknowledged outdoor sport, fitting somewhere between gardening and deer hunting.

And finally, if only to prove that wood heat's day has arrived (again), we're beginning to see the beginning of a whole new area of crime: logjacking. Certain unsavory elements are sneaking their trucks into other folks' woods and picking up cordwood ricks where they were laid out to dry. Other creeps are mounting snowmobiles, riding out to snowed-in summer houses and hunting camps, breaking in and taking about the only item of real value left behind, the heating and cook stoves. I suppose we'll see more and more of this in the future. It may take a while sinking in, but wood heat has become more than the fad or folly of a nostalgic minority. It is big business, a growing factor in our nation's energy picture and, sad to say, an object of thievery. So keep your stove close at hand and lock the woodshed at night.

# APPENDIX B
# Hot Water from the Firebox

Many older wood stoves have ports located at the rear of the firebox that are either used in converting to kerosene or to introduce piping to heat water. Our kitchen range has three of these ports, and a neighbor has spotted an old, copper hot-water tank in a relative's house. If we haven't got the tank operating in time to get photos for this book, you may be sure it is in (and working—hopefully) as you read this.

Under an old-time gravity water-heating system, you operate on the sure assumption that hot water will rise. The tank is set up on legs so as to be higher than the stove. Water heated in the firebox will rise in the tube and cooler water at the bottom of the tank will replace it. The system would be more efficient if a small pump were attached, and perhaps we'll put one on.

Several independent inventors and a few small companies are bringing out devices for producing hot water for washing and cooking and/or for heating the house from a stove or fireplace. One gadget for fireplaces runs tubing halfway up the flue to heat water. Others run coils or a series of S-shaped curves through the fire itself, around the back, top or sides of firebox or hearth, or inside or around the outside of a stovepipe. Any piping that goes near the fire, once again, should be one continuous piece. The inside of your ash bank can get up to 2,000 degrees F. and that's enough to melt most welding bonds. Don't take a chance.

The piping you run inside the fire or flue acts as a heat-exchanger. For a house-heating system, attach each end to pipes that run to standard finned baseboard heaters or braze your own fins on copper tubing if you have the time and equipment. Then all you need is a small pump to push the water along and an expansion tank somewhere in the system. A possible homemade system is illustrated, and more than one company is now selling plans. Just be sure you get a pump made to move hot water and

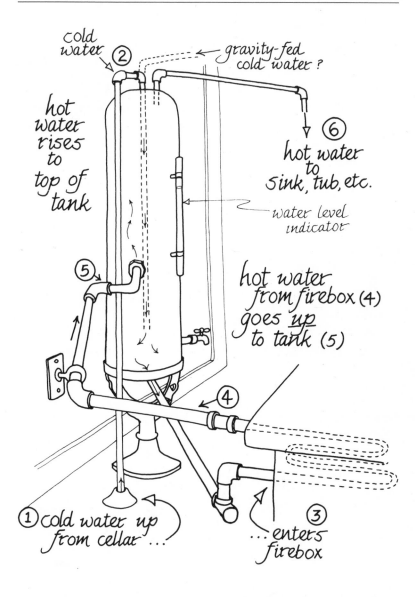

cold water ②

gravity-fed cold water ?

hot water rises to top of tank

⑥ hot water to sink, tub, etc.

water level indicator

hot water from firebox (4) goes *up* to tank (5)

⑤

④

① cold water up from cellar...

③ ...enters firebox

use sturdy copper or iron pipe—not the plastic hot water pipe that is made for domestic hot water systems that seldom get hotter than 140 degrees F. The water coming out of the wood fire may be a lot hotter than that. You'll have to obtain pump, expansion tank and valves from a heating contractor, and he can tell you how to put the system together or will do it for you if needed or if the building code requires.

One thing to remember in building your own water heater or in operating a commercially made unit is that water must be circulating through the pipes whenever you have a fire going. Try pumping cold water into a coil system that's blazing hot, and it will turn to steam and blow the apparatus apart and you with it, likely as not. You can get by with a bit of sputtering and steaming with a little short loop such as we have in the little drum stove I built, as there isn't enough coil to keep the steam confined. But with any sort of closed system carrying water through a good length of pipe in a fireplace or stove, keep the water circulating constantly so long as there is a fire.

Probably the most economical hot water units get their heat from the top of the firebox or stove or fireplace or from the flue or stovepipe. This is using heat that would be lost up the chimney otherwise. The main problem of putting any tubing directly in the smoke is that it will eventually become coated with soot and creosote that may burn and surely will reduce its heat absorptive capacity. Better, I'd think, would be to arrange coils around the outside of a flue liner (or stovepipe as we've al-

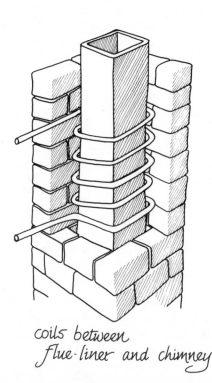

coils between
flue-liner and chimney

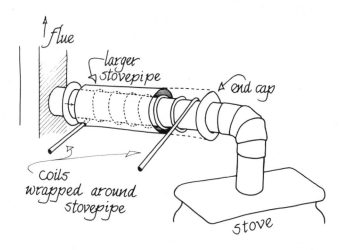

ready suggested). This would be counterproductive in an installation such as our own living room Combi with an outside flue. Superefficient stoves of any design don't let much heat into the flue and what does get out is needed to maintain draft and keep creosote accumulation at a minimum. There's no point cooling it down with a water heater. However, it would be fine for an inside-house flue. If I ever get around to that log cabin with a Rumford fireplace, I'll build a coil into the flue. I will use round flue liner, which is more efficient than the rectangular, and will mortar a coil of copper tubing around the outside of it, in the space between liner and the concrete chimney blocks. Then, I'll run pipes in and out through grooves in the blocks, attach a pump and use the system to heat the upper story of the house. To run heat into distant rooms in the lower story, we'll either run pipe under the hearth (through perforated bricks), or I actually will put in the under-hearth, hot-air blower arrangement mentioned earlier. It is conceivable that a fan could be inserted into the chimney, pushing air through the space between liner and masonry, and out into rooms. If you are the first to do it, let us know.

# BIBLIOGRAPHY

## General Books

Clegg, Peter. *New Low-Cost Sources of Energy for the Home.*
Charlotte, Vt.: Garden Way Publishing Co., 1975.
Gay, Larry. *The Complete Book of Heating With Wood.*
Charlotte, Vt.: Garden Way Publishing Co., 1974.
Glesinger, Egon. *The Coming Age of Wood.*
New York: Simon and Schuster, 1949.
Hand, A. J. *Home Energy How-To.*
New York: Popular Science/Harper & Row, 1977.
Havens, David. *The Woodburners Handbook.*
Brunswick, Me.: Harpswell Press, 1973.
*Home Fires Burning: The History of Domestic Heating and Cooking.*
London: Hillary, 1964.
Kauffman, H. J. *Early American Ironware.*
Rutland, Vt.: C. E. Tuttle, 1966.
Mercer, H. C. *The Bible in Iron.*
Doylestown, Pa.: Bucks County Historical Society, 1941.
Shelton, J. and Shapiro, A. B. *The Woodburners Encyclopedia.*
Waitsfield, Vt.: Vermont Crossroads Press, 1977.
Smith, Elmer, and Horst, Mel. *Early Ironware.*
Lebanon, Pa.: Applied Arts Pub., 1971.
Stoner, Carol H. *Producing Your Own Power.*
Emmaus, Pa.: Rodale Press, 1974.

## Fireplace and Stove Books

Angier, Bradford. *How to Build Your Home in the Woods.*
New York: Hart Publishing Co., 1952.
Coleman, Peter J. *Woodstove Know-how.*
Charlotte, Vt.: Garden Way Publishing Co., 1974.
Curtis, Will and Jane. *Antique Woodstoves: Artistry in Iron.*
Asheville, Me.: Cobblesmith, 1974.
Eastman, Margaret and Wilbur F. *Planning and Building Your Fireplace.*
Charlotte, Vt.: Garden Way Publishing, 1976.
Harrington, Geri. *The Wood-Burning Stove Book.*
New York: Macmillan

Lytle, Marie-Jeanne and R. J. *Book of Successful Fireplaces: How to Build, Decorate and Use Them.* 19th ed.
  Farmington, Mi.: Structures Publisher, 1971.
Merrilees, Douglas. *Curing Smoky Fireplaces.*
  Charlotte, Vt.: Garden Way Publishing, 1973.
Reid, Joe and Peck, John. *Stove Book.*
  New York: St. Martin's Press, Inc., 1977.
Ross, Bob and Carol. *Modern and Classic Woodburning Stoves and the Grass Roots Energy Revival.*
  Woodstock, N.Y.: Overlook Press, Dist. by Viking, 1977.
Schuler, Stanley. *How to Design and Build a Fireplace.*
  New York: Macmillan, 1977.
Sunset Editors. *How to Plan and Build Your Fireplace.*
  Menlo Park, Ca.: Lane Books, 1973.
Vrest, Orton. *The Forgotten Art of Building a Good Fireplace.*
  Dublin, N.H.: Yankee, 1969.
Wik, Ole. *Wood Stoves: How To Make and Use Them.*
  Anchorage, Ak.: Alaska Northwest.
Wigginton, Eliot, ed. *The Foxfire Book. Vol. 1.*
  Garden City, N.Y.: Anchor Books, 1972.

## Cooking Books

Anon. *Simply Good Cooking, Pennsylvania Dutch Style.*
  Comins, Mich.: Mallory-Yoder Publishing, 1977.
Callahan, Hester. *Cast Iron Cookbook.*
  Concord, Calif.: Nitty Gritty Publications, 1969.
Cooper, Jane. *Woodstove Cookery.*
  Charlotte, Vt.: Garden Way Publishing, 1977.
Lyon, J. C. *The Fireplace Cookbook.*
  Santa Fe, N.M.: The Lightning Tree, 1975.
Wigginton, Eliot, ed. *The Foxfire Book.*
  Garden City, N.Y.: Anchor Books, 1972.

## Wood and Logging Books

Allen, Peter H. *Firewood for Heat.*
  Brattleboro, Vt.: Stephen Greene Press, 1974.
Brush, Warren D., and Collingwood, Gilt. rev. ed., *Knowing Your Trees.*
  Washington, D.C.: The American Forestry Association, 1974.
Forbes, Reginald D. *Woodlands for Profit and Pleasure.*
  Washington, D.C.: The American Forestry Association, 1971.
Hall, Walter. *Barnacle Parp's Chain Saw Guide.*
  Emmaus, Pa.: Rodale Press, 1977.
Stephens, Rockwell A. *One Man's Forest.*
  Brattleboro, Vt.: Stephen Greene Press, 1974.
Vivian, John. *How to Select, Cut and Season Good Firewood.*
  Virginia Beach, Va.: Customer Service Dept., Stihl, Inc., 1977.

————. *The Manual of Practical Homesteading.*
  Emmaus, Pa.: Rodale Press, 1975.
Yepsen, Roger B. Jr. *Trees for the Yard, Orchard and Woodlot.*
  Emmaus, Pa.: Rodale Press, 1976.

## U.S. Government Publications: from Superintendent of Documents, U.S. Government Printing Office, Washington, DC 20250

*Logging Farm Wood Crops*, Farmer's Bulletin No. 2090, U.S.D.A.
*Managing the Family Forest*, Farmer's Bulletin No. 2187, U.S.D.A.
*Why T.S.I.: Timber Stand Improvement Increases Profit*, U.S. Forest Service
  PA-901, U.S.D.A.
*Wood Fuel Preparations*, U.S. Forest Service FPL-090, U.S.D.A.
*Public Assistance for Forest Landowners*, U.S. Forest Service PA-893,
  U.S.D.A.
*Fireplaces and Chimneys*, Farmer's Bulletin No. 1889, U.S.D.A.
*Suggestions for the Inexperienced Woodland Buyer*, U.S. Forest Service
  Mimeo Brochure, U.S.D.A.

## Safety

*From National Fire Protection Association, Int'l.*
  *470 Atlantic Avenue, Boston, MA 02210 ($2 each as of Jan. 1976)*
NFPA No. 89m Heat Producing Appliance Clearance, 1971
NFPA No. 10 Extinguishers, Installation, 1973
NFPA No. 25 Rural Water Systems, 1960
NFPA No. 46 Timer: Outside Storage, 1973
NFPA No. 211 Chimneys, Venting System, 1972
NFPA No. 224 Homes Forest Areas, 1972
NFPA No. HS-8 Using Coal and Wood Stoves Safely!, 1974

## Periodicals with Frequent Articles on Wood Heat

*B & K's Country Journal*, P.O. Box 8600, Greenwich, CT 06830
  A glossy monthly devoted to New England country life in the main, but what
  better area for wood heat?
*Farmstead*, P.O. Box 392, Blue Hill, ME 04614
  All aspects of on-the-land self-sufficiency from the state where wood heat
  never became obsolete.
*Harrowsmith*, Camden East, Ontario, Canada K0K 1J0
  A delightful new entry in the self-reliance, small farm field. Glossy, well-re-
  searched and proud to be Canadian.
*Organic Gardening*, Emmaus, PA 18049
  The monthly bible of the natural gardening movement, into living with na-
  ture on all fronts, wood heat stressed.

*Mother Earth News*, P.O. Box 70, Hendersonville, NC 28739
  Thick newsprint bimonthly, all aspects of self-sufficiency, largely reader-written; many new ideas.
*Wood Burning Quarterly & Home Energy Digest*, 8009 Thirty-fourth Avenue South, Minneapolis, MN 55420
  The only magazine I know of devoted largely to wood heat.
*Yankee*, Dublin, NH 03444
  New England local interest magazine; plenty of old-time wisdom, cooking and heating with wood included.

# SOURCES

## Assorted Stove Models

Atlanta Stove Works, Inc., Box 5254, Atlanta, GA 30307
   Full line of antique and modern cooking range designs, radiating and circu-
   lating heaters.
Bow & Arrow Stove Co., Bow & Arrow Streets, Cambridge, MA 02138
   Imports several lines of European cooking and heating stoves; large retail
   outlet.
Cawley/LeMay Stove Company, Inc., Barto, PA 19504.
   U.S.-made cast iron Scandinavian-design stoves.
Dawson Mfg. Co., Shaker Rd., Box 2024, Enfield, CT 06082
   Variety of wood and coal stoves.
Enterprise Foundry Co., Ltd., Sackville, New Brunswick, Canada E0A 3C0
   Modern-design cooking ranges, box stoves. Franklin fireplaces.
Fisher Stove Works, 229 Mill Street, Eugene, OR 97401
   Stair-step design heating/cooking stoves in log and open front styles.
   Welded steel, firebrick lined, cast-iron doors, guaranteed for a lifetime.
HDI Importers, Schoolhouse Farm, Etna, NY 03750
   Imports a variety of European designs.
King Stove and Range Co., Box 730, Sheffield, AL 35660
   Cooking stoves, heaters, laundry stoves.
Kristia Associates, Box 1561, Portland ME 04104
   Importers of total line of stoves, ranges and fireplace designs from Norway.
Maleable Iron Range Co., Beaver Dam, WI 53916
   Monarch line of superior wood, wood/oil, etc. ranges, circulating heaters,
   Franklins.
Markade-Winnwood, Box 11382, Knoxville, TN 37919
   Good barrel stove kits, fireplace and Shaker-style log burners of contempo-
   rary design. Furnace core, in-fireplace stove.
Merry Music Box, 10 McKown St., Boothbay Harbor, ME 04538
   Styria made in Austria; heaters and modern-design cooking ranges and ac-
   cessories. A music box museum too!
Norwegian Stoves, P.O. Box 132, Titusville, N.J. 08560
   Importers of the ULEFOS line of heating and cooking stoves. Traditional
   Scandinavian quality, some highly ornate.

412

"Old Country" Appliances, P.O. Box 330, Vacaville, CA 95688
   Imports the Tirolia line of Austrian cooking (and heating) stoves. Firebrick
   lining, modern clean-line design in decorator colors.
Petzinger's Supply and Service Co., P.O. Box 326, Elizabethtown, KY 42701
   N. American representative of Hass & Sohn, W. German maker of full line
   of heating and cooking stoves—since 1854.
Portland Stove Foundry, 57 Kennebec St., Portland, ME 04104
   Former maker of old stove designs including "Princess" range, now doing
   airtight-baffle type stoves. Damn shame.
Preston Distributing Co., Foot of Whidden Street, Lowell, MA 01852
   Sells full retail line of stoves, imports French-made/Chapee line of automatic
   circulators.
Scandinavian Stoves, Inc., Box 72, Alstead, NH 03602
   Distributors for Lange, a complete line of Danish heating and cooking stoves
   and fireplaces of top quality.
Self Sufficiency Products, One Appletree Square, Bloomington, MN 55420
   Steel stoves, firebrick lined in several sizes; log burner and open models.
Shenandoah Manufacturing Co., Inc., P.O. Box 839, Harrisonburg, VA 22801
   An established maker of circulating stoves, with and without cabinets,
   conventional stoves and accessories such as a blower-equipped, tubular fire-
   place grate.
Southport Stoves, 248 Tolland St., E. Hartford, CT 06108
   The Morso line of Danish cast-iron heating stoves and fireplaces, Esel
   Belgian stoves. Send a dollar for their Wood Heat Handbook of really good
   dope on wood burning.
Thermo-Control Heating Systems, Howe Caverns Rd., Cobelskill, NY 12043
   Variety of stoves, furnaces and accessories.
Timberline International Inc., 5231 Chinden Blvd., Boise, ID 83704
   Steel, firebrick and fabbled step-stoves.
United States Stove Co., P.O. Box 151, S. Pittsburg, TN 37380
   An expanding line of steel and cast-iron heaters; Wonderwood circulator.
Upland Stove Co., Greene, NY 13728
   More copies of Scandinavian designs. Nice castings.
Valco Corp., 215 Johnson Rd., Michigan City, IN 46360
   Mark 1—an improvement on Scandinavian designs.
Warmglow Products, 625 Century, S.W., Grand Rapids, MI 49503
   Copies of Scandinavian designs plus accessories.
Washington Stove Works, Box 687, Everett, WA 98206
   One of the major manufacturers of cast-iron stoves of all kinds. Accessories
   including a 70-pound, three-legged pot for boiling maple sap, witches'
   brews, missionaries and book editors.
Weir Stove Co., Taunton, MA 02780
   One of the few surviving old-time manufacturers of iron stoves.
Wilton Stove Works, 33 Danbury Rd., Wilton, CT 06807
   Several steel stoves of an austere Shaker-type design, a Franklin-type and a
   dandy cooking range in the works.

# Circulating Heating Stoves

Ashley Automatic Heater Co., Box 730, Sheffield, AL 35660
   Circulators in plain and fancy designs.
Atocrat Corp., New Athens, IL 62264
   A front-feed automatic stove (cookstoves too).
Brown Stove Works, Inc., Box 490, Cleveland, TN 37311
   Still another line of circulators.
Foresight Enterprises, Inc., 343 Luinsor St., Ludlow, MA 01056
   The N-ER-G Saver stove. Has a cook-top.
Hunter Enterprises (Orillia, Ltd.), P.O. Box 400, Orillia, Ontario L3V 6K1
   "The Pride of Canada," a thermostatically controlled cabinet heater in
   several sizes.
Locke Stove Co., 114 W. 11th St., Kansas City, MO 64105
   Circulators with side-feed.
Martin Industries, King Products Div., Sheffield, AL 45660
   Full line of automatic circulators with and without cabinets.
Riteway Mfg. Co., Div. Sarco Corp., Box 6, Harrisonburg, VA 22801
   Several models, claims to burn all smoke and gases.
Suburban, P.O. Box 399, Dayton, TN 37321
   Both wood and coal circulators.

# Central Heating Units

Arotek Corp., 1703 East Main., Torrington, CT 06790
   The Hoval offers both central heat and domestic hot water in a wood/coal or
   wood/oil furnace.
Bellway Mfg., Grafton, VT 05146
   Custom-built wood, wood and oil, conversions in hot air or hot water
   furnaces.
Charmaster Products, Inc., Grand Rapids, MN 55744
   Makes the Charmaster wood and oil furnace.
Dualfuel Products, Inc., Box 514, Simcoe, Ontario, Canada N3Y 4L5
   Oil/wood combination hot air furnaces.
Duo-Matic of Canada, Waterford, Ontario, Canada N0E 1Y0
   Oil, wood, coal-burning hot air furnaces, the Newmac.
F & W Econoheat, Inc., Star Route, Box 53A, Bozeman, MT 59715
   The Boyd Heating & Ventilation System: hot air wood furnace fits into fire-
   place, uses outside air for combustion.
Fawcett, Sackville, N.B. E0A 3C0 Canada
   Furnacaid main or auxilliary wood furnace.
Integrated Thermal Systems, 379 State St., Portsmouth, N.H. 03801
   Distributes the Danish-made HS Tarm wood/electricity, oil or gas central
   hot water home heating and domestic water supply furnace.
Johnson Energy Systems, Inc., 7350 N. 76 St., Milwaukee, WI 53223
   Auxilliary wood stove for warm air furnace. "$500 will handle".
Kikapoo Stove Works, Box 127, La Farge, WI 54639
   A stove maker that also manufactures the Home Furnace central heater.

Longwood Furnaces Corp., Gallatin, MO 64640
  Wood/gas or oil, forced hot water furnace will hold wood up to five feet long.
Marathon Heater Co., Inc., Box 167, R.D. 2, Marathon, NY 13803
  A logwood burner capable of 200,000 Btu's.
Northern Heating & Sheet Metal, Grand Rapids, MN 55744
  The Combo, a wood/oil combination furnace.
Ram Forge, Brooks, ME 04921
  Wood-burning supplements to your hot water or hot air furnace, guaranteed
  25 years.
Riteway Mfg., Div. Sarco Corp., Box 6, Harrisonburg, VA 22801
  Makes line of wood- and oil- or coal-burning hot air and water furnaces
  (stoves too).
S/A Energy Division, Crossroads Park Dr., Liverpool, NY 13088
  Imported wood/coal, gas/oil furnace units home hot water capability.
Thermo-Control Wood Stoves, Cobleskill, NY 12043
  Makes wood stoves to supplement existing hot air or hot water central heat-
  ing systems.
Wilson Industries, 2296 Wycliff, St. Paul, MN 55114
  The Yukon wood/oil or wood/gas hot air furnace.

# Miscellaneous and Single-Design Stoves

All Nighter Stove Works, Inc., 35 Nutmeg Lane, Glastonbury, CT 06033
  Several sizes of a stair-step design steel stove with cast-iron door, tubing
  through firebox for blower-forced air circulation.
American Stovalator, Rt. 7, Arlington, VT 05250
  An in-fireplace air-circulating stove.
Arizona Forest Supply, Box 188, 610 E. Butler Ave., Flagstaff, AZ 86002
  A blown, baffled circulating steel design. Burns "buffalo chips" too.
Boston Wind Power Co., 574 Boston Ave., Medford, MA 02155
  Sells the Independence, a steel stove, baffled and a modification of Scandi-
  navian design by Larry Gary, author of *The Complete Book of Heating With
  Wood*. A soapstone-clad circulator too.
Buffalo Stoves International, Box 5526, Incline, NV 87450
  A great hulking thing with four iron doors.
Chimney Heat Reclaimer Corp., 53 Railroad Ave., Southington, CT 06489
  A stovepipe heat saver and the "carrot top," forced hot-air heating/radiating
  stove.
Country Craftsmen, P.O. Box 3333, Santa Rosa, CA 95402
  A reasonably priced kit to convert steel drum to stove.
Earth Stove, 10725 S.W. Tualatin-Sherwood Rd., Tualatin, OR 97062
  Thermostatic fireplace steel stove. Ornate.
Freeflow Stove Works, So. Strafford, VT 05070
  Three sizes of stove made from curved tubing to heat and circulate room air
  more rapidly.
Garden Way Catalog, 1300 Ethan Allen Ave., Winooski, VT 05405
  Sells the Reginald 101, an Ireland-made copy of the small Scandinavian
  airtights and others sent knocked down, by mail.

General Products Corp., Inc., 655 Plains Rd., Milford, CT 06460
Tubular grates and blowers.

Hayes-te Equipment Corp., Unionville, CT 06085 and C and D Distributors, Inc., Box 715, Old Saybrook, CT 06475
Sellers of the "Better 'N' Ben's" box stove and accessories—stove vents into existing fireplaces.

Hinkley Foundry & Marine, 13 Water St., Newmarket, NH 03857
Sand mold, hand cast iron stoves in traditional shaker austere style. Lovely.

Hurricane Stove., Rt. 1, Box 154, Pingree, ID 83262
A steel stove. Water reservoir available too.

Isotherimics, Inc., 291 River Rd., Clifton, NJ 07014
Makes heat pipes for heat-saving, in-stove pipe installations.

J and J Enterprises, 4065 W. 11th Ave., Eugene, OR 97402
Another Oregon step-stove. 12-point design patent for what good that will do you.

Kikapoo Stove Works, Box 127, La Farge, WI 54639
Makes steel stoves with refractory liner, cast doors in an unusual design.

K.N.T., Inc., Box 25, Hayesville, OH 44838
"The Impression," a modern Franklin design with forced-air circulating feature. Comes in black or colored tile.

Landau Metal Products Corp., Newman and Ross Inc., 250 5th Ave., New York, NY 10001
"The Kent" stove/fireplace combination.

Metal Building Products, 35 Progress Ave., Nashua, NY 03060
Nashua stove combines traditional steel box stove design with a manifold for circulating blown hot air.

Mohawk Industries, Inc., 173 Howland Ave., Adams, MA 01220
A top-loading, downdraft stove of modern design, The Tempwood.

New Hampshire Wood Stoves, Inc., Box 310, Plymouth, NH 03264
"Home Warmer" baffled steel stove.

R and K Incinerator Co., Rt. 4, Decatur, IN 46733
Burn-Easy Lifetime camp stove, a small drum-style stove ready to light.

Ridgeway Steel-Hydroheat Div., Box 382, Ridgway, PA 15853
Several stove and fireplace designs that heat water or air for central heat.

SEVCA Stove Works, Box 396, Bellows Falls, VT 05101
Drum-type stove with secondary heat/combustion chamber main drum from used propane tanks, baffled; lots of heat, reasonable.

Thermo Matic, Rt. 145, Lawyersville Rd., Cobleskill, NY 12043
Steel step-stoves in several sizes.

Vermont Castings, Inc., Box 126, Prince Street, Randolph, VT 05060
The Defiant in two sizes. Cast-iron stove made in U.S.A. Airtight, baffled, burn open or closed, Franklin fireplace-type design.

Vermont Soapstove Co., Inc., Perkinsville, VT 05151
As of this report (Jan. 1978) the only manufacturer of soapstove griddles. No soapstove wood stoves available at this date; maybe they are making them now. Please send a stamped, self-addressed postcard if you inquire. A small company.

Vermont Techniques, Inc., Box 107, Northfield, VT 05663
A dual-walled air circulating stove to fit into an existing fireplace.

Vermont Wood Stove Co., 307 Elm St., Bennington, VT 05201
The "Downdrafter" claims to burn all combustion gases. They will send you a good little book: "Buyer's Guide to Wood Stoves" which isn't as biased toward their design as it could be.

Yankee Wood Stoves, Cross St., Bennington, NH 03442
Drum stoves, assembled and inexpensive.

# Fireplace-Type Stoves and Equipment

Aqua Appliances, Inc., 13501 Sunshine Ln., San Marcos, CA 92069
Circulating stove that fits into existing fireplace with blowers and glass doors.

Ecology Heating, Inc., 805 Main St., Farmington, NM 87401
Hot water furnace that fits into your existing fireplace.

Energy Centers of America, Inc., 5276 Valley Industrial Blvd., S. Shakipee, MN 55379
A forced-air fireplace heat reclaimer that fits in the firebox well above the flames. Must be built into the brickwork.

Fire-View Distributors, P.O. Box 370, Rogue River, OR 97537
The Fire-View wood heater in four sizes. A patented design with tempered glass window for viewing fire.

Garden Way Research, Charlotte, VT 05445
Patented grate. By mail only.

Greenbriar Products, Box 473, Spring Green, WI 53588
A modern-design free-standing fireplace, semi-airtight stove. Accessories for hot-water and hot-air control heating conversion.

Hubbard Creek Trading Co., Star Rt., Box 82, Silverton, OR 97381
Heat Saver brand heat saver.

Leyden Energy Conservation Corp., Leyden, MA 01337
The Fuel Mizer, a free-standing or fireplace insert circulating stove with glass doors.

Loeffler Heating Units Mfg. Co., R.D. 1, Box 503-0, Branchville, NJ 07826
Another tubular grate manufacturer.

Majestic Co., Hungtington, IN 46750
Another prejob fireplace maker. Dampers, and ventilating brick.

National Sewer Pipe, Ltd., Box 1800, Oakville, Ontario, Canada L6J 4Z4
Ceramic chimney pots.

Radiant Grate, Inc., 31 Morgan Park, Linton, CT 06413
The Dahlquist's grate and cooking accessories.

Superior Fireplace Co., 21325 Artesia Ave., Fullerton, CA 92633
Heatforth warm-air circulating fireplaces.

Texas Fireframe Co., P.O. Box 3425, Austin, TX 78764
Provides the "Physicist's Fire," grate designed by a PhD. Opens up coals, permitting burning of unsplit wood.

Thermograte Enterprises, 51 Lona Lane, St. Paul, MN 55117
Tubular grate.

Vega Industries, Inc., Mt. Pleasant, IA 52641
Make a Heatilator fireplace unit and dampers.

Vestal Manufacturing Co., Box 420, Sweetwater, TN 37874
Dampers, etc.

Winnwood Industries, 4200 Birmingham Rd., Kansas City, KS 64117
Circulating Franklin-style stove to be inserted in an existing fireplace. Also steel stoves, furnace core.

# Mail Order Suppliers with Catalogs

L. L. Bean, Freeport, ME 04032
Stove offerings in catalog change frequently. Retail store in Freeport. A major supplier of outdoor gear of all sorts.

Countryside Catalog, 312 Portland Rd., Waterloo, WI 53594
Cast-iron cookware, barrel stove kit, ranges, small but select stove selections including the patented Fisher steel stoves.

Cumberland General Store, Rt. 3, Crossville, TN 38555
Most every wood stove design there is, cast-iron ware, accessories such as a galvanized wash boiler.

Edmund Scientific Co., 555 Edcorp Bldg., Barrington, NJ 08008
Many hobby items including heat savers.

Enterprise Sales, Apple Creek, OH 44606
A $1 catalog of full line of heating and cooking stoves.

Glen-Bel's Country Store, Rt. 5, Box 390, Crossville, TN 38555
$3 gets you a catalog of self-sufficiency items, complete line of stoves and accessories, books and etc.

W. F. Landers Co., Box 211, Springfield, MA 10010
Perhaps the largest collection of ranges, stoves and accessories.

Lehman Hardware & Appliances, Inc., Box 41, Kidron, OH 44636
"Survival center" of the Midwest, kerosene lamps, farm supplies and a full line of stoves and accessories.

Mother's General Store, Box 506, Flat Rock, NC 28731
Jotul, all kinds of U.S.-made stoves, cooking ranges, accessories, cast-iron cookware and more.

NASCO Farm and Ranch Catalog, NASCO, Ft. Atkinson, WI 53538
or NASCO, 1524 Princeton Ave., Modesto, CA 95352
No stoves, but a good mail-order source of logging equipment.

Pioneer Lamps & Stoves, 71J Yesler Way, Pioneer Sq. Stn., Seattle, WA 98104
Claims to have the nation's largest line of new and old stove designs. $3 for catalog.

Retsal Corp., McCammon, ID 83250
Small but good selection of self-sufficiency items including their own cookstoves, barrel stove kits, fireplaces, pipe and etc.

# Accessories

Jaxco, Inc., Wautoma, WI 54982
  The handiest fire tool there is: a pair of insulated horsehide gloves that let
  you pick up red-hot coals if you like.
Jefferson Boat Works, Box 931, Jefferson, MA 01522
  A cordwood sawhorse.
Louisville Tin and Stove Co., Box 1079, Louisville, KY 40201
  Stovepipe drum oven (the "Progress," sootless and with thermometer).
Mechanic Street Exchange, R.D. 1, Mechanic Street, Galway, NY 12074
  Stove mica in most every size you'd warm up to.
Patented Mfg. Co., Bedford Rd., Lincoln, MA 01773
  "Slip-On Heat Fins" for stovepipes.
Peavey Manufacturing Co., Box 371, Brewer, ME 04412
  Right, they make a peavey for about $20 mail order.
Thermal Reclaimation, 939 Chicopee St., Chicopee, MA 01021
  Pipe-located heat saver. Thermo-saver.
Vermont Crafts Market, Main St., Box 17, Putney, VT 05346
  Mail order catalog with the best fireplace tools I've seen, handmade and
  reasonably priced.
Worcester Brush Company, Box 658, Worcester, MA 01601
  A full line of chimney sweep gear: brushes, rods and fittings.

# Wood Splitters and Saws

## Hydraulic Types

Albright Corp., Jeffersonville, VT 05464
  Tractor, self-powered. Components and plans.
Amerind-MacKissic, Inc., Box 111, Parkerford, Pa. 19457; the Mighty Mac.
Didier Manufacturing Co., Box 163, 8630 Industrial Drive, Franksville, WI
53126; the Hydra-Splitter.
Futura Enterprises, West Bend, WI 53095
Huss Sales & Service, Toledo, OH 43601
Lindig Splitters, 1875 W. County Road, St. Paul, MN
Piqua Engineering, Inc., Box 605, Piqua, OH 45356; the Lickity Splitter.
Servotrol, Inc., Box 48, Monroe, GA 30655
  Small, relatively cheap unit.

## Screw Types

Derby Splitter, Box 21, Rumson, NJ 07760
  Self-powered portable. Tractor PTO-driven unit
FW and Associates, Inc., 1855 Airport Road, Mansfield, OH 44903
  Self-powered Bark Buster
Knotty Wood Splitter Co., Route 66, Hebron, CT 06248
  Tractor PTO-driven

Nortech Corporation, Midland Park, NJ 07432
  Self-powered
Taos Equipment Manufacturers, Box 1565, Taos, NM 87571; the Stickler.
Thackery Co., Columbus, OH 43216
  Wheel-mounted Unicorn
Trans America Power Equipment, Inc., 8308 Washington St., Chagrin Falls, OH 44022
  Self-powered Quick Split

## Miscellaneous

C & D Distributors, Inc., Dept. O, Box 766, Old Saybrook, CT 06475
  Upright guided-wedge Jiffy Woodsplitter
Knotty Wood Splitter Co., Route 66, Hebron, CT 06248
  Tractor-driven cordwood saw.
Richard Pokrandt Mfg., RD. 3, Box 182, Tamaqua, PA 18252
  Tractor mounted combination cordwood saw and wood splitter
Thrust Manufacturing, Inc., 6901 S. Yosemite, Englewood, CO 80110

# INDEX

Illustrations appear in boldface.

## A

## B

# C